Table of contents:

I0071499

This book is review of engineering research methods, gives an introduction into the background studied of various research areas; such as agriculture, chemical engineering, electrical engineering, manufacturing and industrial engineering. There are extensive reviews for PhD work. The articles will help those interested in research and head start in their work.

Main editor: *Dr. Nor Mariah Adam*
Co-editor: *Dr. Shahryar Sorooshian*

Chapter 1: Methodology Process of Predictive Maintenance Development through Condition Monitoring at SiNx Deposition Process

S. Nurhaiza [1], M.K.A. Ariffin [2], NM Adam [3]

Department of Mechanical Engineering & Manufacturing Engineering, Faculty of Engineering, UPM

Email: [1]nurhaiza.shahrir@yahoo.com, [2] khairol@eng.upm.edu.my, [3]mariah@eng.upm.edu.my

ABSTRACT

This paper analyzes failures at Silicon Nitride (SiNx) deposition equipments in a solar cell manufacturing through gathering back historical maintenance pass-down logbook. This paper starts with prioritizing the frequent failures through Pareto Analysis and then discuss on potential method to reduce its failure. Time Based Maintenance has already being implemented and researching further towards developing Predictive Maintenance through Condition Monitoring to reduce the equipment frequent failures. This deposition equipment stores most parameter and process readings that makes the automatic Condition Monitoring is possible. A preliminary result show a possible reduction of 35% on one of the frequent failure after the implementation.

Keywords: predictive maintenance (PdM) , maintenance strategy ,condition base maintenance, condition monitoring

1. INTRODUCTION

There are 16 Silica Nitride deposition equipment in this solar cell manufacturing process. Each equipment consist of 4 processing tubes. The raw material for this process is silicon wafer. The wafers are deposited with a chemical gas (Silicon Nitride) to create an antireflective coating on the silicon wafer . The Silicon Nitride reduces the reflection from 30% to 10%. This process is done only after the Silicon wafers are being P-doped and Edge isolated .

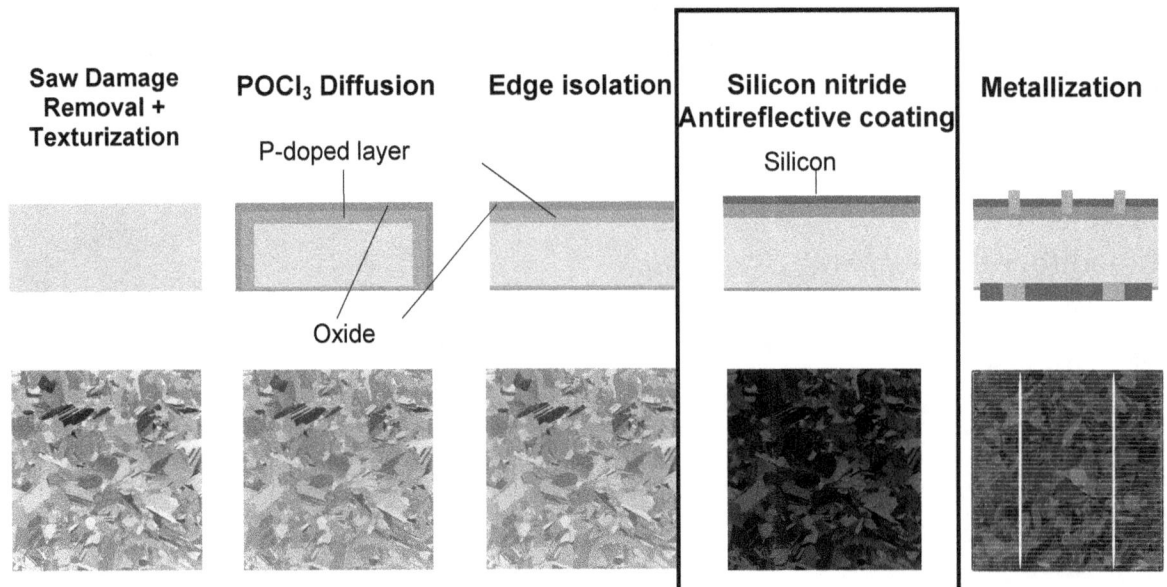

Fig. 1 : Solar Cell Basic Manufacturing steps

There are 4 Quartz tube for each of SiNx Deposition Equipment. Each Quartz tube assembly is as in Fig 2.

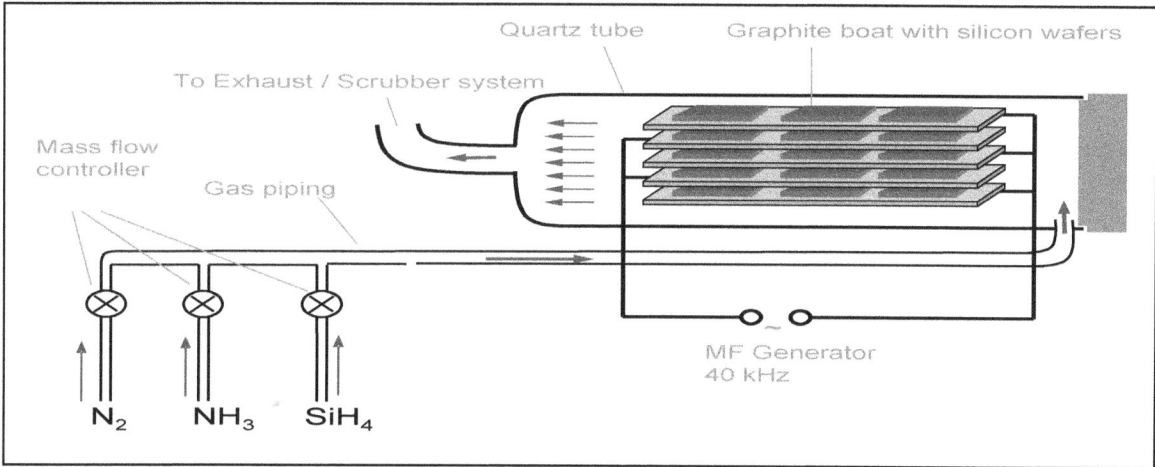

Fig . 2 : 1 Quartz tube assembly in SiN Deposition Equipment

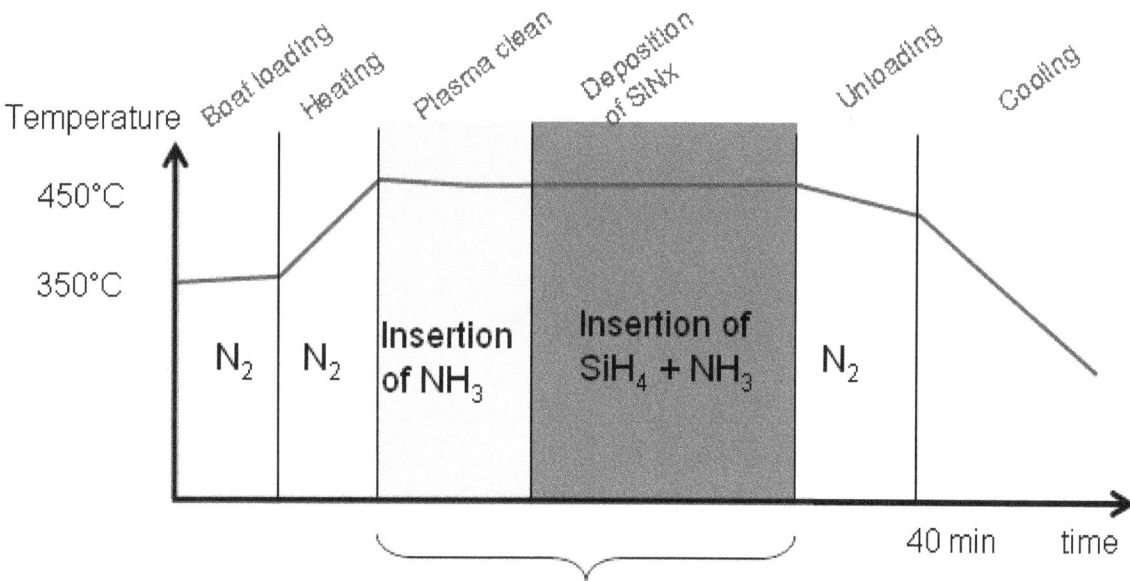

Graph 1. Show the Steps of SiNx Deposition process for a tube.

The flow of N2 , NH_3 and SiH_4 are control by the Mass Flow Controller to make the SiNx deposition to the silicon wafers. Graph 1 shows the steps of SiNx deposition process in a tube. The waste gases will be sucked out from the Exhaust into the Scrubber system which is the Burner Wash (CTBW) (Refer Fig 3).

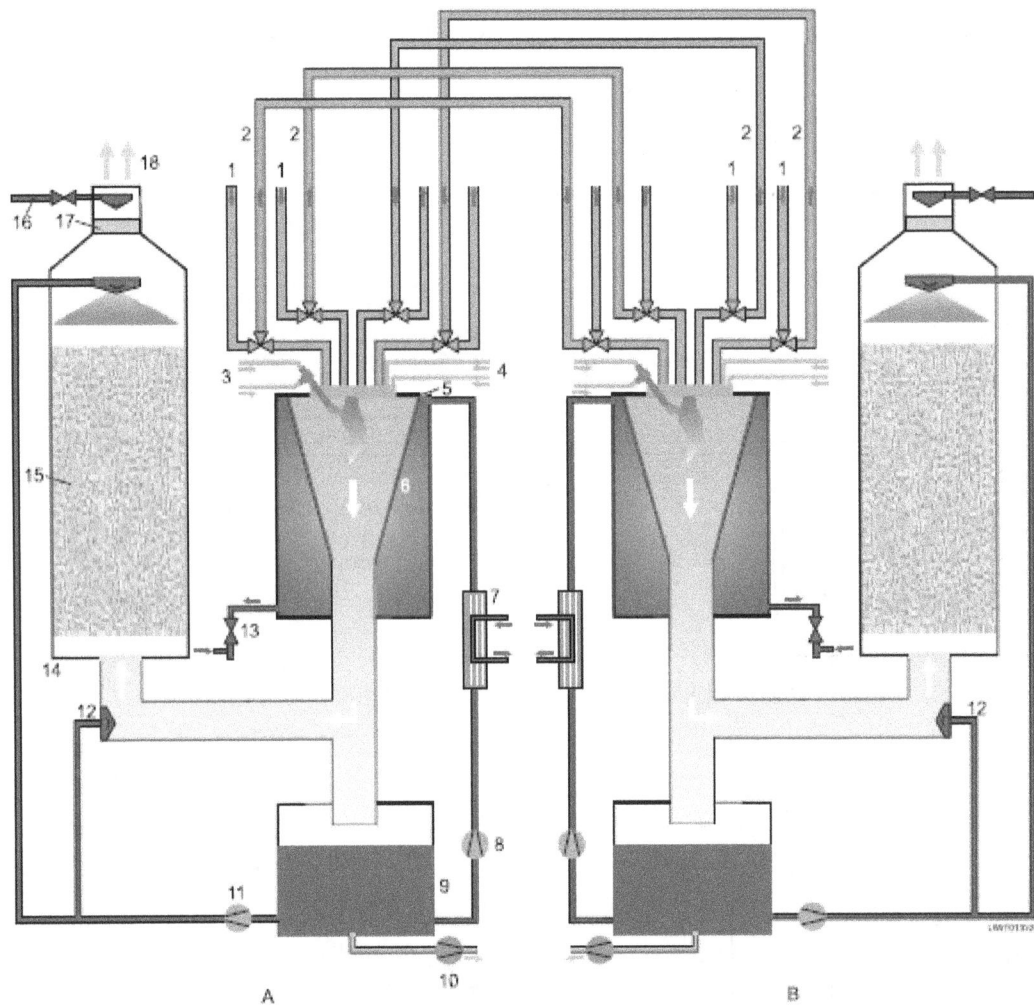

1	Waste gas inlet	10	Waste water depletion pump
2	Bypass	11	Washing liquid circulation pump, scrubber
3	Pilot burner gas supply	12	Washing liquid spray nozzle
4	Main burner gas supply	13	Drain reactor unit
5	Spill level of washing liquid	14	Scrubber column
6	Reactor chamber	15	Filling material
7	Heat exchanger	16	Fresh water spray nozzle
8	Washing liquid circulation pump, reactor	17	Demister
9	Washing liquid tank	18	Clean gas outlet

Fig . 3 : Burner Wash

There is already Time Base Maintenance which are weekly, monthly, quarterly, semi annual & yearly. Maintenance department has already followed all the recommendation from equipment vendor and other plant on requirement of the schedule maintenance.

2. LITERATURE REVIEW

Silicon nitride (SiNx) is a promising material for anti-reflection coating and passivation of multicrystalline silicon (mc-Si) solar cells (Kim et. al,2003). In this work, a plasma-enhanced chemical-vapor deposition (PECVD) system with batch-type reactor tube was used to prepare highly robust SiNx films for screen-printed mc-Si solar cells (Kim et. al,2003).

A hydrogen-rich SiNx layer was deposited using a PECVD system manufactured by Centrotherm (Kim et. al,2003). The horizontal tube furnace for PECVD had a capacity of 72 wafers in a batch for 12.5 by 12.5 cm mc-Si wafers (Kim et. al,2003). The deposition temperature was set in the range between 400 - 450 °C (Kim et. al,2003).

The Preventive Maintenance does not take the actual machine health into consideration, unnecessary machine shutdown might be inevitable, which may incur unnecessary cost (Junhong et. al, 2005). Predictive Maintenance is a maintenance policy in which
selected physical parameters associated with an operating machine are sensored, measured and recorded intermittently or
continuously for the purpose of reducing, analyzing, comparing and displaying the data and information obtained for
support decisions related to the operation and maintenance of the machine (Rao, 1996).

Based on Hawaiian Electric Company, it has demonstrated that predictive maintenance system can be deployed and have a major impact on reducing unplanned outages (Emoto et. al, 2006).

The key benefits of effective equipment Condition Monitoring system includes :

- Early detection of equipment defects to prevent failures, especially those catastrophic failures which might lead to extensive power interruptions.
- Improved maintenance practices (in coordination with Reliability-Centred Maintenance) – reduction in maintenance frequency, outage duration and maintenance cost.
- Establishment of equipment performances database to facilitate trend analysis to identify equipment type problem.
- Supporting the evaluation of "health indices" and "asset lives" to facilitate the long term financial planning preparing for asset refurbishment or replacement.

(Tang, 2009)

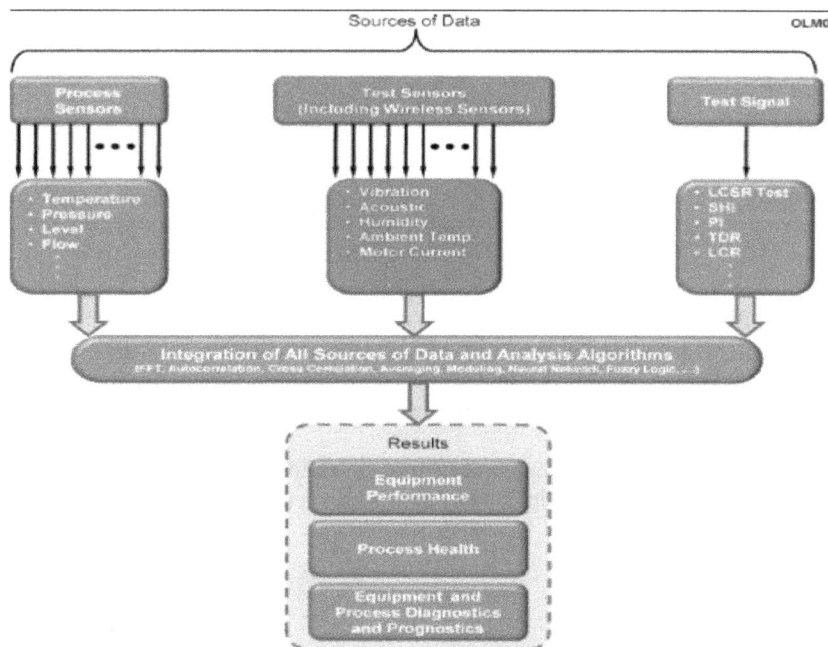

Fig 4. Integrated system employing three technique for predictive maintenance (Hashemian, 2011)

Based on H.M. Hashemian, there are 3 sources of Data for predictive maintenance which are existing sensor-based maintenance techniques, test sensor-based maintenance techniques and test signal based maintenance techniques.

Industrial plants should no longer assume that equipment failures will only occur after some fixed amount of time in service; they should deploy predictive and online strategies that assume any failure can occur at any time (randomly)
(Hashemian, 2011).

The presence of a Computerized Maintenance Management System (CMMS) in an industrial plant has been considered a previous requirement for the implementation of a Predictive Maintenance Programmes (Carnero, 2005). The Predictive Maintenance Programmes require software and systems for signal acquisition to obtain information (Mobley, 2001). Therefore, in order to implement a Predictive Maintenance Programmes (PMP), it is considered necessary for the personnel of the maintenance department to be computer literate and therefore to be able to use and maintain the predictive equipment and the information it provides (Carnero, 2005).

3. PRIORITIZATION FOR CONDITION MONITORING

The equipments' overall downtime performance is as Figure 5.

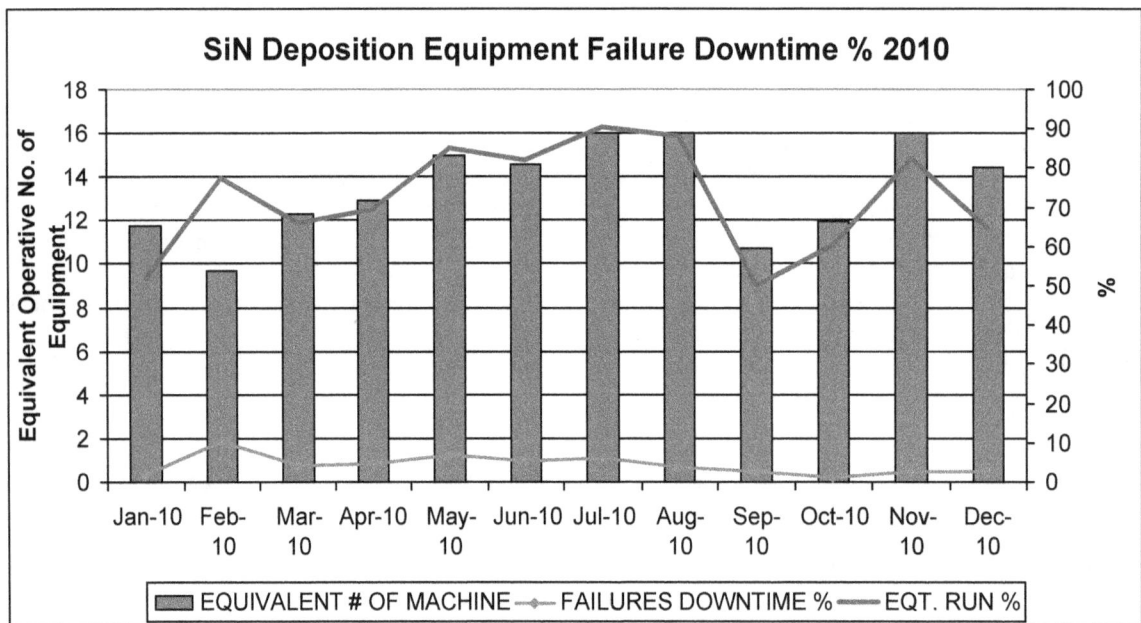

Fig 5. Failure Downtime % of Deposition Equipment in year 2010

The Pareto analysis which is also known as 80–20 rule, is named after the Italian economist Vilfredo Pareto (Surhone et. al, 2010). The principle states that for many events, roughly 80% of the effects/problems come from 20% of causes (Jayswal et. al, 2011). The Pareto analysis helps focusing the attention on the most important causes instead of wasting time and energy on minor ones (Jayswal et. al, 2011).

The team also have extracted the yearly maintenance pass-down to analyze the frequent failure and have constructed into the Failure Pareto as in Fig 5. The Pareto chart is also summarizing the failure frequency which is shown downtime % in Fig 6.

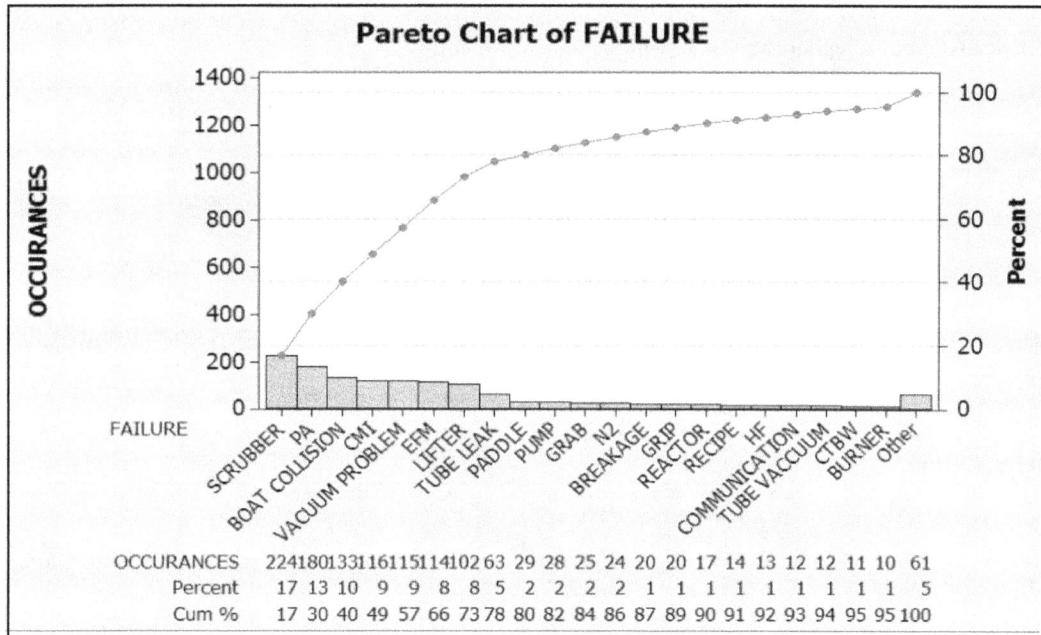

Figure 6 . SiN$_x$ Deposition Equipment Failures Pareto in year 2010

Based on the Failure Pareto, Scrubber & PA (Process Abort) are the two highest failure occurrences. The team concluded that in order to further reduce the failures in future is by focusing on the Scrubber and Process Abort preventive solution.

Before implementing the Condition Monitoring, a Root Cause Analysis need to be done. Root cause analysis is an approach used to identify the reason of a problem. It can be done by an individual or group (Motschman and Moore, 1999). The more complex the problem and involved the process, the greater the need to enlist a team of individuals and formalize the analysis (Motschman and Moore, 1999). The team should include people who: perform the steps of the process, oversee the process, are knowledgeable about the process but are not directly involved in its performance or oversight, and are knowledgeable about process improvement activities and tools (Motschman and Moore, 1999).

Table 2 presents as an example of the possible causes for absorber tower corrosion incidents with probability ratio (which a lower number shows a higher probability) .

Table 2: Most likely root causes for absorber tower corrosion incidents (Harjac et. al, 2007)

Probability	Potential root cause	Description
1L	Inadequate V^{5+} concentration	No magnetite. Insufficient V^{5+} accelerates corrosion when a break in a sulfide / hydrocarbon layer occurs.
2A	Hydrocarbon accumulation in top bed	Prevents magnetite and/or sulfide layer formation by poor wetting, or may inhibit corrosion until disturbed
2A	High raw gas loading	Mechanical damage of protective layer/scale. May sweep hydrocarbons upward thus preventing passive layer establishment in upper region.
2A	Frothing/Foaming	Mechanical depassivation Chemical action: locally low pH areas accelerate attack.
3A	Galvanic	Galvanic interaction between dissimilar sulfide and steel, leading to accelerated dissolution.
3A	Formation of sulfide layer due to presence of H_2S	Sulfide layers contradict the anodic inhibition management programme which is designed to pursue a magnetite layer
	Inadequate H_2S concentration may prevent adequate sulfide coverage	Sulfide layer thickness decreases with height in the absorber tower, suggesting a problem with the protection by a sulfide mechanism at the absorber top

Latent (L) and active (A) causes are distinguished.

Similarly , before starting to implement a predictive maintenance, the most frequent potential root causes for Scrubber & Process Aborts to happen at Silica Nitrate (SiNx) Deposition equipment need to be known.

Maintenance team had list the potential root causes with some ratios on probability based on the team's experience whereby a higher ratio is showing a higher probability.

Table 3. Root Cause Analysis Table with Probability Ratio

Failure	Potential Causes	Current Prevention	Ratio	Actions Recommended
Scrubber	Condensation build up within time internally at Mass Flow Controller(MFC) which causes MFC efficiency drop	i.Centralise LPG filter to minimize condensation ii.Cleaning activity every week	90	i. Install Strainer ii. Predictive Maintenance by analyzing Condition Monitoring of Mass Flow Control.
	scrubber door is not close	Alarm detecting door not fully close	10	
Process Abort	Alarm sihi pump trigger due to high negative vacuum pressure/vacuum leak occur.	Adjust water and gas condition	40	Predictive Maintenance using Condition Monitoring of Leak Test.
	Pre-Deposition steps took longer	Broken wafers are found inside tube	10	Predictive Maintenance using Condition Monitoring of Heat up time.
		Wrong feedback by the thermocouple to the controller.	20	Predictive Maintenance using Condition Monitoring of Heat up time.
		SiNx Valve is not working or not open/close fully.	10	Predictive Maintenance using Condition Monitoring of Butterfly Valve.
		Soft Pump working too long than normal	10	Predictive Maintenance using Condition Monitoring of Soft Pump.

Based on the Table 3, the top probability root causes for Scrubber is the condensation build up within time internally in the Mass Flow Controller (MFC). The team have concluded to analyze the Condition Monitoring of the Mass Flow Control performances but unfortunately in actual, these data is not available. To measure these parameter, it requires installation of PLC system and requires cost to be spent. Therefore, this action was de-prioritize. But for Process Abort , there are 4 potential condition monitoring that can be used for predictive maintenance and these readings are already pre-stored in the machine, and only require a schedule extraction into a database for monitoring. From the existing stored readings, the team can study the readings pattern and decide to take actions based on abnormality. Below Fig. 7 is the methodology on how to implement a Predictive Maintenance using Condition Monitoring.

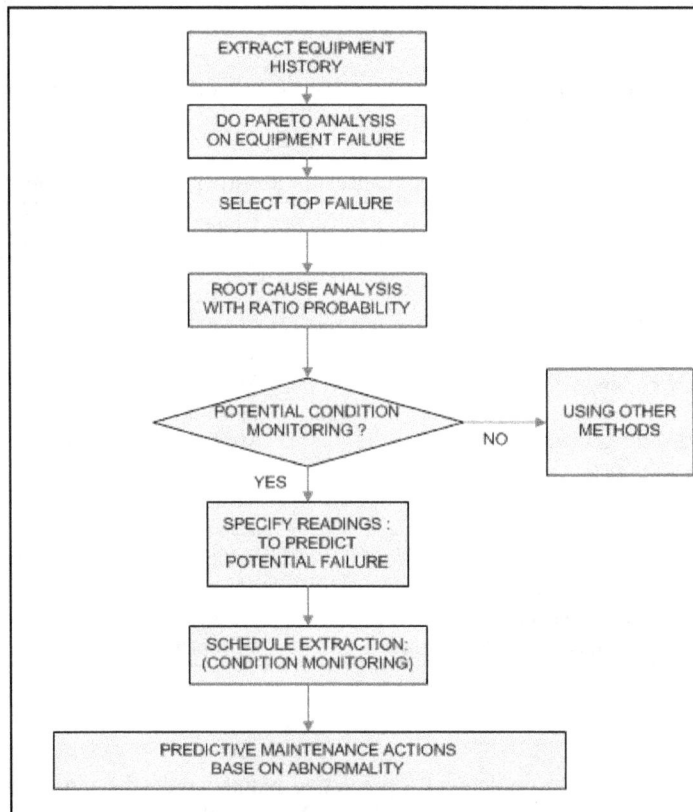

Fig 7. Methodology of Predictive Maintenance using Condition Monitoring at Silicon Nitride (SiNx) deposition equipment

5. CONCLUSION

This research paper have concluded the methodology for developing the Predictive Maintenance using the Condition Monitoring
of a SINx deposition equipment to reduce frequent failure. It starts with a Pareto Analysis on the failures of the SINx deposition equipment and performing Root Cause analysis on the top failures with probability ratio based on team's experience. From the Root Cause analysis, the team will list if there a potential condition monitoring and schedule a data extraction for condition monitoring on pre-stored data. Predictive Maintenance actions will be taken based on the abnormality. Preliminary result shows a reduced Process Aborts failure after the implementation of Heat Time & Leak Test Condition Monitoring in May 10 and May 12, 2011 (Refer Table 4) . The Process Aborts reduced from average 17.82 Process Aborts / running equipment / month to 11.61 Process aborts / running equipment / month which is a reduction of 34.8%. The other implementation of the whole Condition Monitoring of Butterfly Valve & Soft Pump for this SiNx deposition equipment will be discussed on the next paper.

Table 4. Comparison of average Process Abort (PA) frequency with and without Predictive Maintenance
by Condition Monitoring .

DURATION	PA PROCESS ABORTS (PA) (TOTAL FREQUENCY)	AVERAGE MONTHLY /# OF EQUIVALENT EQUIPMENT RUNNING	AVERAGE PA/MONTH/ RUNNING EQUIPMENT
JAN 2010- DEC 2010 (without Condition Monitoring)	180	10.10	17.82
MAY 2011 – SEPT 2011 (with Heat Time & Leak Test Condition Monitoring)	36	3.10	11.61

REFERENCES

[1] Junhong Zhou, Xiang Li, Anton J.R. Andernroomer, Hao Zeng, Kiah Mok Goh, Yoke San Wong and Geok Soon Hong (2005). Intelligent Prediction Monitoring System for Predictive Maintenance in Manufacturing. Industrial Electronics Society, 2005. IECON 2005. 31st Annual Conference of IEEE Page 2314 - 2319 .

[2] Clesson T. Emoto, Rudy Tamayo, and Gary R. Hoffman, Senior Member IEEE (2006). Implementation of a Predictive Maintenance System. Transmission and Distribution Conference and Exhibition, 2005/2006 IEEE PES Page 57 - 61 .

[3] Joe, C.Y. Tang, (2006).CLP Experience on Condition Monitoring and Condition Based Maintenance. Advances in Power System Control, Operation and Management (APSCOM 2009), 8th International Conference Page 1 - 7 .

[4] H.M. Hashemian, Senior Member, IEEE. State-of-the-Art Predictive Maintenance Techniques. IEEE Transaction on Instrumentation and Measurement 2010. Volume 60, Issue 1, Page 226 - 236.

[5] J.Kim , J. Hong and Soon Hong Lee . Application of PECVD SINx films to Screen Printed Multicrystalline Sillicon Solar Cell. Journal of the Korean Physical Society (2004), Volume 44, No.2, Page 479-482

[6] Rao BKN. Handbook of condition monitoring. Oxford: Elsevier; 1996.

[7] Mobley K. Plant engineer's handbook. Reading, MA: Butterworth- Heinemann; 2001.

[8] M. Carmen Carnero. An evaluation system of the setting up of predictive maintenance programmes. Reliability Engineering and System Safety 91 (2006) 945–963.

[9] Surhone, L., Timpledon, M., & Marseken, S.. Pareto analysis: Statistics, decision making, Pareto principle, fault tree analysis, failure mode and effects analysis. Pareto distribution. Wikipedia Betascript Publishing 2010.

[10] Abhishek Jayswal, Xiang Li, Anand Zanwar, Helen H. Lou and Yinlun Huang (2011). A sustainability root cause analysis methodology and its application. Computers and Chemical Engineering *35 (2011) 2786– 2798*

[11] Tania L. Motschman and S. Breanndan Moore, 1999. Corrective and preventive action. Transfusion Science 21 (1999) 163-178.

[12] S.J. Harjac a, A. Atrens a, C.J. Moss. Six Sigma review of root causes of corrosion incidents in hot potassium carbonate acid gas removal plant. Engineering Failure Analysis 15 (2008) 480–496

Chapter 2: Progress and Current Trends on Modified Homogeneous Charge Compression Ignition (HCCI) from SI (Spark Ignited) Engines Using EGR Method

By: Ahsanul Kaiser,
M.Sc. in Mechanical Engineering
Faculty of Engineering, UPM

ABSTRACT

Homogeneous charge compression ignition (HCCI) combustion has been implemented because of high thermal efficiency and lower nitrogen oxide (NOx) and particulate matter emissions to meet the demand of fuel economy and greenhouse effect. Although, still facing the challenges in the perfect successful operation of HCCI engines at the combustion control, cold start, limited power output and also unburned hydrocarbon(HC) and CO emissions. It has been found that exhaust gas recirculation (EGR) has the significant potential to control the HCCI combustion and also increase the operating range to high loads, and also the most important path to get gasoline - fuelled HCCI combustion. The current strategies are described in the paper on Spark Ignition (SI) to modify HCCI engines.

Keywords: HCCI, EGR, SI, HC & CO.

Abbreviations

AFR	air fuel ratio		
ATDC	after top dead centre	IMEP	indicated mean effective pressure
BDC	bottom dead centre	LTR	low temperature reactions
BTDC	before top dead centre	NOx	nitric oxides and nitrogen dioxides
BMEP	brake mean effective pressure	NVO	negative valve overlap
CAI	controlled auto-ignition	PCCI	premixed charge compression ignition
DI	direct injection	PCI	premixed compression-Ignited
EGR	exhaust gas recirculation	PM	particulate matter
EVC	exhaust valve closed	TDC	top dead centre
HC	hydrocarbons	SI	spark ignition
HCCI	Homogeneous Charge Compression Ignition	UNIBUS	uniform bulky combustion system
ID	ignition delay		
IEGR	Internal Exhaust Gas Recirculation		

Content

1. Introduction

2. Background of HCCI

Introduction

In the 21st century, Green House Gas emissions and fuel consumption are two vital worldwide environmental and energy challenges. Homogeneous Charge Compression Ignition (HCCI) is nonetheless combustion process that have advantages both highly efficient and low NOx and particulate matter emissions, because can provide high, diesel like efficiencies using gasoline, diesel fuel, and most alternative fuels with ultra-low emissions of NOx and particulate matter and In some regards, HCCI includes the best features of both Spark Ignition (SI or Gasoline) engines and Compression Ignition (CI or Diesel) Engine. So this system can be a promising combustion method in Internal Combustion Engine to improve the thermal efficiency of motor vehicle in order to establish an environmentally friendly, sustainable society [14].

Background of HCCI
Gasoline Engine & Diesel Engine

The Gasoline engine as shown in Fig. 1.1, has been recognized as one of the most promising powertrain due to its low emissions and reliability. Modern emission standards and customer demands for less consumption have increased the need for advanced design theories for Gasoline engines, also it has the large heat losses and pumping losses. The diesel engine as shown in figure 1.1 , has been recognized as the most promising powertrain of the foreseeable future due to its superior thermal efficiency and reliability [1]. Modern emission standards and customer demands have increased the need for advanced design theories for diesel engines. The result is more performance with improved fuel efficiency. A tiny quantity of fuel is injected into the chamber and ignited before the primary combustion takes place. This minimises the harsh "knocking" or rattle sound traditionally associated with diesel power plants [2].

HCCI Fundamentals:

In HCCI engines, a uniform charge of air and fuel enters the combustion chamber, then the compression stroke gives sufficient energy input to the mixture, so that the temperature reaches at an auto-ignition condition ,so then the charge (of a homogeneous mixture) auto-ignited in the combustion process. This combustion process without the help of any external source of ignition, unlike conventional engines where ignition is started with either injected fuel or spark respectively. So an HCCI can be told as a hybrid of SI (Gasoline) engine and CI (Diesel)engine, as because in HCCI engine fuel is homogenously premixed with air(SI engines),but the fuel auto-ignites from compression heating as in CI engines. In such a way, homogeneous air- fuel mixture auto-ignites at many locations as shown in fig 1.1.

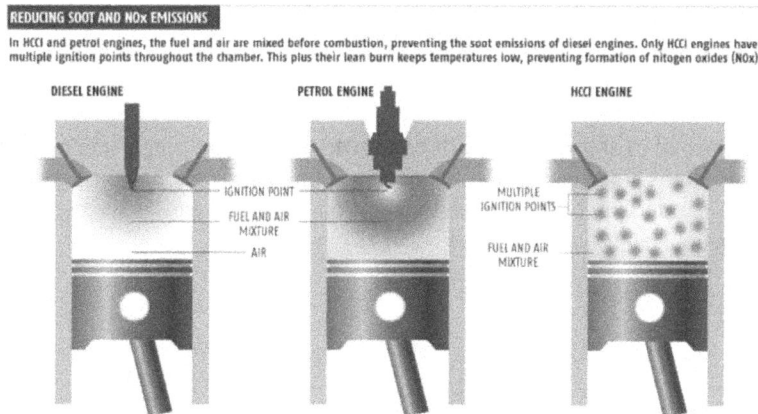

Fig. 1.1

The mixture is homogeneous which minimises emission of NOx and PM, it is compression ignited using high compression ratios, has no throttling losses, also has shorter combustion duration which leads to high thermal efficiency.
On the other hand the SI principle has low efficiency at part load where HCCI is better cause good efficiency, also the CI(Diesel) engine has the similar efficiency quality as the HCCI, but it produces higher amount of PM and NOx.

HCCI depends on the auto-ignition of highly diluted or lean air-fuel ratio mixture which allows an HCCI engine to run with the advantage of less fuel consumption compared to Spark or Diesel engines. Low temperature combustion due to the presence of excess air and diluents suppresses the rate of NOx formation, see the below figure 1.2 [30]. In this way HCCI is a promising attractive concept for combustion engines to reduce both emissions and fuel consumption at low temperature.

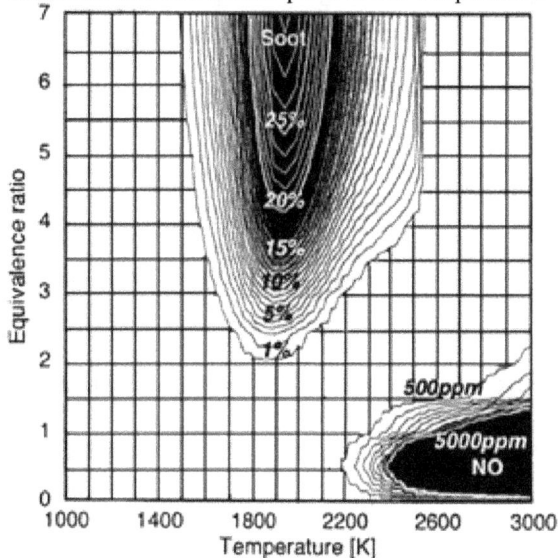

Fig. 1.2 Equivalence ratio vs Temperature

Application of HCCI

The application of HCCI engines has become major interest of automotive industry since the advantages of this particular engine can improve engine efficiency by producing low NOx and soot emissions according to previous results from [6] and [7]. In fact, HCCI technology could be scaled to virtually every size-class of transportation engines from small motorcycle to large ship engines. HCCI is also applicable to piston engines used outside the transportation sector such as those used for electrical power generation and pipeline pumping. HCCI engines are particularly well suited to series hybrid vehicle applications because the engine can be optimized for operation over a more limited range of speeds and loads compared to primary engines used with conventional vehicles [8].

The analysis and experimental as shown by [9] will be needed to study the single and multi-zone combustion of the HCCI system taking the full account of the temperature gradient in the combustion chamber. Perhaps this system will be able to produce a better fuel efficiency and cleaner emissions by improving the ignition performance and combustion control as shown in Fig. 2 [10].

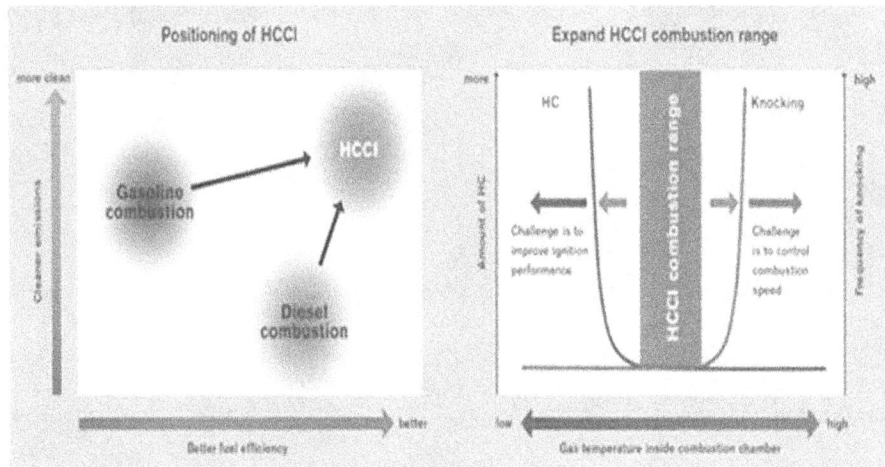

Figure 2

HCCI History

Although it is known as a new combustion concept for internal combustion engines in many papers , HCCI has also known as Controlled Auto Ignition(CAI) has been around over hundred years. The first patent refers to invent a Hot Bulb 2 stroke oil engine by Carl W. Wesis in 1897[13]. Then the Russian scientist Nikolai Semonov and his colleagues established the first theoretical and practical exploitation of chemical kinetics controlled combustion for diesel engines in the 1930s [14] and later in the 1970s by Gussak [16] who built the CAI engine that controls combustion by using active species which are discharged from partially burned mixture in a separate prechamber.

But the most recognized and first systematic early investigations on CAI were done individually by Onishi [6] and also Noguchi [11] on two stroke engines in the late seventies at 1979.Then on a four stroke engine by Najt and Foster in 1983[7]. Also 'the terminology Homogeneous Charge Compression Ignition (HCCI)' was introduced by Thring[12] in his research paper on studying the effect of external EGR and air-fuel ratio on HCCI .

Limitation of HCCI

Control of Combustion Timing, Limited Power Output, Homogeneous Mixture Preparation and Cold start.

Methods of HCCI

There is no direct means to control HCCI, the several indirect control methods which influence HCCI combustion are given below:

- Variable Exhaust Gas recirculation by externally
- Negative Valve Overlapping to control EGR internally
- Supercharging
- Variable compression ratio
- Mixing a per oxide additive to improve ignition property and
- Adopting duel fuels with different ignition property
- Introducing a laser to stimulate HCCI ignition
- Air pre-heating
- Mixing a per oxide additive to improve ignition property and
- Adopting duel fuels with different ignition properties

Why EGR?

HCCI Still is limited to engine laboratories, even almost 30 years passed after HCCI was invented and demonstrated in 4-stroke engines. The important ways of controlling HCCI combustion by variable exhaust gas recirculation externally and internally by negative valve overlapping (NVO). EGR is needed at the specific strategy of fuel modification and the addition of EGR into suction /intake which helps to increase the temperature further during the compression stroke of the next cycle as previously explained [15], is the more practical way of controlling charge temperature in an HCCI engine rather than the existence of of active radicals. The hot inserted gases contained in the EGR easily can be used to control the heat release rate because of its impact on chemical reaction which can delay the auto ignition timing and lower peak cylinder pressure.

Another reason to delay in the introduction of HCCI combustion is that the strategy can be applied across limited load range.

Control strategy of gasoline fuelled at HCCI using EGR

At high loads there is insufficient dilution of the charge, the chemical reactions are too fast and the combustion is too rapid, which results in pressure waves travelling back and forth in the combustion chamber. These pressure waves are manifested as pressure fluctuations in in- cylinder pressure traces, similar to knock in SI combustion, but a more appropriate name is 'ringing' [19]. One way to avoid this ringing phenomenon is by boosting to allow more air to be taken in and the charge to be diluted more, as investigated in references [20] and [21]. The problem with boosting is finding a turbocharger (apart from mechanically driven compressors) that can both be viably operated at the typically low exhaust energies associated with HCCI and deliver high boost pressures in high-load SI operation. Sjo¨ berg and Dec et al [22] showed that using fuel stratification was another effective method to smooth the heat release and reduce the pressure-rise rate, in tests with a primary reference fuel exhibiting two- stage ignition. Further, although Leach et al. [23] used NVO for combustion control. His study has also shown that the rate of heat release can be decreased by using a less homogeneous mixture than in standard HCCI operation where gasoline fuel was only used.

Other control methods which related to EGR to control ignition

Xie et al. [24] investigated the influence of spark ignition on CAI combustion based on internal EGR strategy in a single engine. The results show that spark ignition can play an important role in controlling CAI combustion ignition in low load boundary region. The low temperature chemical reaction process is shortened and the auto-ignition timing is advanced due to the spark discharge. Meantime, lower fuel consumption and cycle-to-cycle variations can be achieved. The spark discharge makes auto-ignition easier in low load boundary region, so that the CAI operation can be expanded to lower load through assisted spark ignition. , it has the potential of spark ignition for control of auto-ignition timing and control of CAI/SI mode switch. Hyvonen et al. [27] investigated the spark assisted HCCI combustion during combustion mode transfer to SI in a multi-cylinder VCR-HCCI engine. They found that spark assisted can be used for controlling the combustion phasing during a mode transfer between HCCI and SI combustion. However, the combustion fluctuations are large in the intermediate combustion region where some cycles have both spark ignited flame propagation and auto-ignition, but some cycles have only partially burnt flame propagation. More articles can be found by Berntsson et al[28], Felsch et al [26], Bunting et al [25]

NVO and external EGR are applied to across limited power

Another reason for delays in the introduction of HCCI combustion is that the strategy also can only be applied across a limited load range.

Cairns and Blaxill et al [17] using in his research a combination of internal and external EGR has been used to increase the attainable load in a multi-cylinder engine operated in gasoline controlled auto ignition. The amount of residual gas trapped in the cylinder was adjusted via the NVO method. The flow of externally re-circulated exhaust gas was varied due to use a typical production level valve. Under stoichiometric fuelling conditions, the highest output achieved using internal exhaust gas was limited by excessive pressure rise and unacceptable levels of knock. Introducing additional external exhaust gas was found to retard ignition, reduce the rate of heat release and limit the peak knocking pressure. In Fig. 3, it can be seen that addition of external EGR enabled significant increase in peak engine output, rising from 350 kPa to 580 kPa (~65%). This supplementary cold diluting gas served to retard ignition, evident in the phasing of the 10% mass fraction burned (Fig. 4).

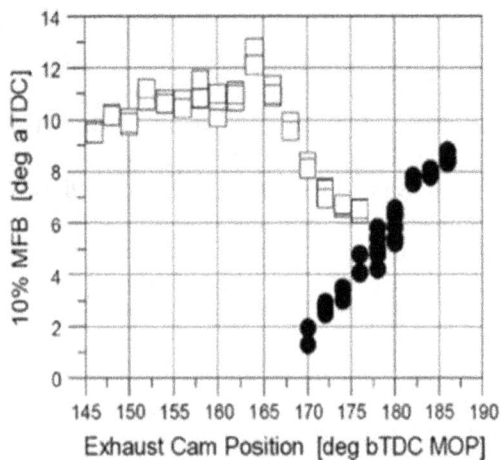

Fig 4 : 10% mass fraction burned versus exhaust cam position [244].

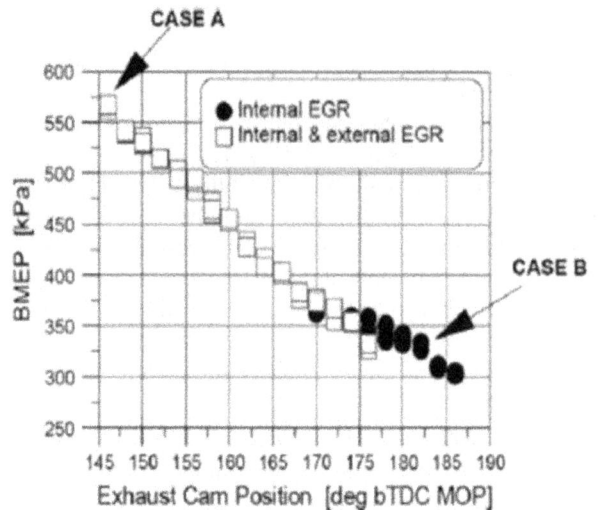

Fig 3 : BMEP versus exhaust cam position (1500 rpm, $\lambda = 1.0$) [244].

A continuation of work presented by Dahl et al [18], where it was shown that it is possible to use a stratified charge to reduce the maximum pressure- rise rates and ringing intensities in an HCCI engine with NVO. The stratified charge was created by adding a late injection, in the later part of the compression stroke, to the main injection in the early part of the intake stroke, hence the combustion strategy was called stratified charge compression ignition (SCCI). The engine was run at 1200 r/min and 4 bar IMEP (indicated mean defective pressure) with a constant combustion phasing. It was found that using charge stratification was effective for reducing ringing when using both a primary reference fuel that displayed two-stage ignition and a gasoline that showed little or no two-stage ignition. With 50 per cent of the fuel injected late (end of injection (EOI) = 30 crank angle degrees (CAD) before top dead centre (BTDC)) ringing intensities were reduced by about two-thirds in operation with both fuels, and the maximum pressure-rise rate was reduced to half the values observed in homogeneous combustion. EGR and NVO also used at more articles to increase load which can be found by Dahl et al [29].

Conclusion

HCCI is a combustion concept which has developed over the years in response to the need for lower NOx and soot emissions with higher efficiency increasing from SI gasoline and CI diesel engines. Hence, for future long term development of HCCI combustion systems, the key issues will be more flexible injection strategies and EGR control for better mixture formation and control as well as high boost to extend the upper load limits. With projected increasing flexibility in both engines hard- ware and control system in the long term, the development of a full HCCI engine is possible. Nevertheless, progress has been made within the HCCI automotive field to overcome this as evident by the commercial applications of the UNIBUS by Toyota Motor Corporation, the MK combustion system by Nissan Motor Corporation, General Motor and some others. In the United States, General Motors Company has prioritised HCCI technology and is anticipating the introduction of HCCI engines to the market by 2012 [31]. In conclusion, whilst HCCI remains a realistic alternative to existing engine combustion technologies to improve emissions, it should continue to have long term viability as an energy source for both light-duty and heavy-duty vehicles.

References

[1] Xin, Q. Diesel Engine System Design, Cambridge: Woodhead Publishing Limited, 2011.

[2]Castrol limited, diesel vs petrol http://www.castrol.com/castrol/genericarticle.do? 1999-2011.

[3] http://static.howstuffworks.com/gif/diesel-two-stroke.gif

[4] Westbrook, C.K. The Internal Combustion Engine at Work. Online at https://www.llnl.gov/str/Westbrook.html , 2010.

[5] http://c0002954.cdn2.cloudfiles.rackspacecloud.com/blog/wp-content/uploads/2009/03/hcci- explained.jpg

[6] S. Onishi, S. H. Jo, P. D. Jo and K. Shoda. SAE Paper no. 790501, 1979.

[7] P. M. Najt and D. E. Foster . Compression-Ignited Homogeneous Charge Combustion. SAE Paper no. 830264, 1983.

[8] U.S. Department of Energy, Energy Efficiency and Renewable Energy Office of Transportation Technologies. Homogeneous Charge Compression Ignition (HCCI) Technology, A Report to the U.S, Congress, 2001.

[9] Aceves, S. M., Flowers, D. L., Martinaz-Frias, J., Ray Smith J., Dibble, R., Au, M. and Girrard J., HCCI Combustion: Analysis and Experiments, SAE Technical Paper Series, 2001.

[10] http://www.nissan-global.com/EN/TECHNOLOGY/OVERVIEW/hcci.html

[11] M. Naguchi. " A study on Gasoline Engine Combustion by observation of reactive products." SAE Paper, 1979.

[12] R. H. Thring. "Homogeneous Charge Compression Ignition(HCCI) Engines". SAE Paper no. , 1989.

[13] O. Erlandsson. "Early Swedish Hot-Bulb Engines - efficiency and performance compared to contemporary Gasoline and Diesel engines." SAE Paper no. 2002-01-0115, 2002.

[14] European Commission, Reducing emissions from light duty vehicles. Available from http://ec.europa.eu/environment/air/transport

[15] (i.)Willand, J., Nieberding, R. -G. The knocking syndrome- its cure and its potential. SAE Paper no. 982483, 1998 and

(ii.)Denbratt , I. Method of controlling the process of combustion in an IC engine with means for controlling the engine valves. Volvo Car Corporation, Sweden, patent no. SE21782, 1998 .

[16] L.A. Gussak, SAE Paper No. 750890, 1975.

[17] Cairns, A and Blaxhill, H. SAE Paper no. 2005-01-0133, 2005.

[18] Dahl,d., Andersson ,M., Berntsson, A., Denbratt,I., and Koopmans, L. SAE Paper no. 2009-01-1785, 2009

[19] Eng, J. A. Characterization of pressure waves in HCCI combustion. SAE Paper no. 2002-01-2859, 2002.

[20] Yap, D., Wyszynski, M. L., Megaritis, A., and Xu, H. Applying boosting to gasoline HCCI operation with residual gas trapping. SAE Paper no. 2005-01-2121,2005.

[21] Johansson, T., Johansson, B., Tunesta°l, P., and Aulin, H. HCCI operating range in a turbo-charged multi cylinder engine with VVT and spray-guided DI. SAE Paper no. 2009-01-0494, 2009.

[22] Sjo¨berg, M. and Dec, J. E. Smoothing HCCI heat- release rates using partial fuel stratification with two-stage ignition fuels. SAE Paper no. 2006-01-0629,2006.

[23] Leach, B., Zhao, H., Li, Y., and Ma, T. Control of CAI combustion through injection timing in a GDI engine with an air-assisted injector. SAE Paper no. 2005-0134, 2005.

[24] Xie H., Yang L., Qin J., Gao R., Zhu H. G., He B. Q. The effect of spark ignition on the CAI combustion operation. SAE Paper no. 2005-01-3738; 2005.

[25] Bunting B G. Combustion, control and fuel effects in a spark assisted HCCI engine equipped with variable valve timing. SAE Paper no. 2005-01-0872, 2005.

[26] Felsch C., Sloane T. , Han .J, Barths H. and Lippert A . Numerical investigation of recompression and fuel reforming in SIDI- HCCI engine. SAE Paper no. 2007-01-1878, 2007.

[27] Hyvonen J,Haraldsson G, Johansson B . Operating condition using spark assisted HCCI combustion during combustion mode transfer to SI in a multi-cylinder VCR HCCI engine. SAE Paper no. 2005-01-0109, 2005.

[28] Berntsson A W, Andersson M, Dahl D, Denbratt I . A LIF study of OH in the negative valve overlap of a spark assisted HCCI combustion engine. SAE Paper no. 2008-01-0037, 2008 .

[29] Dahl ,D. , Denbratt, I., HCCI/SCCI load limits and stoichiometric operation in a multi cylinder naturally aspirated spark ignition engine operated on gasoline and E85 published at 13th October, 2010.

[30] Kitamura T., Ito T., Kitamura Y., Ueda M., Senda J. and Fujimoto H. Soot kinetic modelling and empirical validation on smokeless diesel combustion with oxygenated fuels. SAE Paper no. 2003-01-1789, 2003.

[31] Frost & Sullivan Automotive Practice. Homogenous charge compression ignition (HCCI) - Holy grail of future internal combustion engines. <http:// www.frost.com/prod/servlet/market-insight-top.pag?docid=178537741> [accessed 03.09.10]

[32] Mingfa Yao; Zhaolei Zheng; Haifeng Liu . Progress and recent trends in HCCI engines . Elsevier at Progress in Energy and combustion science 35 (2009) 398-437 at 2009.

[33] Suyin Gan, Hoon Kiat Ng b, Kar Mun Pang b . Homogeneous Charge Compression Ignition (HCCI) combustion: Implementation and effects on pollutants in direct injection diesel engines . Published Elsevier Applied Energy 88 at 2011.

Chapter 3: A review of methodologies for determination of Physical properties of seeds

Bande, Y.M*, Adam, N.M*, Azmi, Y. ** and Jamarie, O**

* Department of Mechanical and Manufacturing Engineering
** Department of Biological and Agricultural Engineering
University Putra Malaysia

43400, Sri Serdang
Sri Kembangan
Selangor Darul Ehsan
Malaysia

Abstract

Comprehensive review on the methodologies for determination of physical properties of seeds is presented. Several physical properties are responsible for design of a processing system, and all are related to the level of moisture content of seeds. In this paper, some of the most important methodologies are briefly discussed. Conclusions of many authors were presented on sub-headings. Conclusions of most authors were that all physical properties increase or decrease with moisture content.

Keywords: moisture content, repose angle, hardness and toughness, sphericity, dimension, surface area, porosity, density, seed

1.0 Introduction

Seeds of agricultural products have been and will continue to be the major source of food worldwide (Babasaheb, 2004). The plant family, Poaceae contributes more food than any other family, which is a seeded grasses or cereals, followed by Fabaceae, constituting legumes and pulses, such as groundnuts, peas and beans. Cereals provide most importantly, carbohydrates and some proteins, while legumes and pulses are richer in proteins. Oilseeds like soybeans, melon, cotton seeds, sunflower, groundnut, safflower, rapeseed and palm are the major constituent sources of edible oil and proteins. Seeds from the fruits and vegetables are important source of vitamins and minerals. In addition, seeds are used in spices, make beverages and drugs. To cap it all, seeds from agricultural crops are used in preparation of fiber, paints, vanishes, soaps, detergents and so many other products for man's daily use.

With the current trend in technological advancement, agricultural seeds (and some farm produce, mostly by-products) are now used as a source of fuel, called the bio-fuel. Among the most celebrated seeds for the bio-fuel are jatropha seeds (Garnayak, et al, 2008), niger seeds (Solomon and Zewdu, 2009), melon seeds (Solomon et al, 2010), Soybean (Azmi and Stephen, 1994) and others.

Seed dimensions (width, thickness and length), 1000 seed mass, surface area, porosity, terminal velocity, sphericity, static coefficient of friction against different materials, repose angle, hardness, true and bulk densities are among the properties that are considered "physical" by researchers as per seed. These variables are central in classification of seeds and in design of machines for processing seeds. They are central because they give technical information to the designer on what are the basic "inputs" for their design.

One of the most important steps in undertaking any of the physical properties test is the cleaning process. All seeds due for these analyses need to be cleaned to obtain a good result, which will reflect the information about the seed. These may include; dust, dirt, stones and chaff and even immature and broken seeds (Bulent et al., 2006; Garnayak et al., 2008; Solomon and Zewdu, 2009). In addition to cleaning, other operations such as grading and initial moisture contents are measured by appropriate mechanisms and in some seeds, grinding and heating may be necessary (Dutta et al., 1988; Altuntase et al., 2005; Aviara et al., 2005).

2.0 Initial Moisture content

The initial moisture content is determined by subjecting seed to oven drying (either air or vacuum) at temperatures between 80°C to 110°C for 8hrs to 24hrs by ASAE standards (S.352.3, 1994). The moisture content is then determined on either wet or dry basis. This approach has been implemented by many researchers. To prepare the seeds for other tests, certain amount of water is added based on weight of sample, initial and final moisture contents of the seeds. The relation below is used (Mohsenin, 1970) in Sacilik et al, 2003) is as follows.

$$Q = \frac{W_i(M_f - M_i)}{100 - M_f} \qquad\qquad 1$$

Q	=	mass of water added (kg)
W_i	=	initial mass of sample (kg)

M_i =   initial moisture content (% d.w)

M_f =   Final moisture content (% d.w)

Before the actual tests are conducted, the seeds are placed in polyethylene bags/plastic containers and sealed, then refrigerated at 5°C for between few hours to a week to allow for proper distribution of moisture in the seeds. The required quantity of the seeds for experiment are then removed from the refrigerator and allowed to reach equilibrium with room temperature for two to three hours, as adopted by Singh and Goswami (1998) and Ozarslan (2002). This is a process by which the actual moisture content of seed is determined. It forms the base line data for all tests to be carried out. References will be made from the value of initial moisture content. Garnayak et al (2008) after cleaning, determine the initial content by using standard hot air oven method on *jartropha curcas L* seeds. They obtained 4.75 % dry basis. In the work of Mustafa (2007) on barbunia bean seed, the initial moisture content was 18.33% dry basis after oven drying for 24hours. Yelcin et al., (2007) on pea seeds started their experiment on pea seeds at 10.06% dry basis moisture content. The final moisture content in their experiment was 35.08%. For sweet corn, the initial content of 11.54% d.b was used and experiments were conducted to 19.74% as reported by Bulent et al., (2006). The procedure of determining the initial moisture content is adopted by many researchers considering different seeds, such as Sharma et al (2011) on tung oil seed from 12.76% to 14.76% d.b, Ogunjimi et al., (2002) on locust bean seed at 10.25% d.b, Vanesa et al., (2008) on chia (*Salvia hispanica*) seeds was 7.00% d.b, Yelcin and Ersan (2007) on coriander (*Coriandrum sativum L*) seeds in the range of 7.10% to 18.94% d.b, Omobuwajo et al (2000) on sorrel (*Hibiscus sabdariffa*)seeds where single value of 7.7% w.b initial moisture content was used for all measurements. Omobuwajo et al., (2003) on calabash (*monodora myristica*) seeds used 7.67% wet basis to determine the physical properties. Shkelqim and Joachim (2010) on jatropha seeds and kernels, Erica et al., (2006) on safflower was 6.9% dry basis, Ebubekir et al., (2005) on fenugreek (*Trigonella foenum-graceum*) seeds was 8.9% to 20.1% dry basis.

3.0   Seed Dimensions

The linear seed dimensions are the length, width and thickness. They are determined for any seed, using a Vanier caliper or micrometer screw gauge, preferably digital. The seed is placed carefully between the grips of the micrometer screw gauge. It is gently screwed to tighten until the gauge clips and the digital reading is fixed. This is repeated in a suitable number of replicates for all the required measurements to reduce error. The simple relations used to determine the arithmetic and geometric diameters utilizes the values of linear dimension. These values are used as follows (Mohsenin, 1970).

$$D_a = \frac{L+W+T}{3} \qquad\qquad 2$$

$$D_g = (LWT)^{1/3} \qquad\qquad 3$$

D_a =   Arithmetic mean

D_g =   Geometric diameter

The findings of Garnayak et al (2008) on jatropha seeds concluded that dimensions of seeds are directly proportional to the level moisture content. However, the sphericity of 0.66 did not change with moisture content from 4.75% to 12.16%. Mustafa (2007) worked on the barbunia seeds and the dimension of seeds obtained were 16.68 mm, 9.36 mm and 7.51mm for length, width and thickness respectively at moisture content of 18.33%. In the work of Yelcin *et al.,* (2007) for pea seeds, the moisture content has tremendously affected the seed dimensions. Yelcin and Ozarlan (2004) on vetch concluded that within the range of moisture content of 10.57% to 20.63%, the dimensions have increased by not less than 15%. Sacilik *et al* (2003) reported on hemp and Ogut (1998) on white lupin. Bulent *et al.,* (2006) reported that the sweet corn seed's size was affected by the moisture positively. This report was similar to Dursun *et al.,* (2007) on sugerbeet, where the dimensions increased by not less than 10% with the moisture content rise from 8.4 to 14.0%. Solomon and Zewdu (2009) on niger seed similarly observed that the dimensions increased by about 11% with the rise of moisture from 5.60 to 31.67%, Oyelade *et al.,* (2005) on African star apple, Omobuwajo *et al.,* (2000) on ackee apple (*Blighia sapida*) Paksoy and Aydin (2004) on edible squash (*Cucubita pepo L*), Dursun and Dursun (2005) on caper seeds all reported similar trend in dimensions with moisture content change.

4.0 Sphericity

Sphericity is the degree of closeness of seed to a sphere. It describes the rolling ability of seed during processing. The sphericity of any seed is a function of the basic dimensions (length, width and thickness). Equation 4 gives the relationship. Garnayak *et al* (2008) calculated the sphericity on jatropha seeds individually, using the geometric and major axis of the seed. However, relating to other research findings, it was concluded that a value of sphericity of 0.7 – 0.8 only is considered spherical for a seed (Dutta *et al.,* 1988; Bal and Mishra, 1988). They concluded that *Jatropha* is not spherical in shape. The sphericity of barbunia bean seed increased from 0.632 to 0.644 with increase in moisture content, as reported by Mustafa (2007). Similar reports have been presented by Aydin *et al.,* (2002) for Turkish mahaleb, Sahoo and Srivatava (2002) on okra seed and Sacilik *et al.,* (2003) on hemp seeds.

$$\emptyset = \frac{(LWT)^{1/3}}{L} \qquad\qquad 4$$

The sphericity increased as reported by several researchers, is similar to that found in sweet corn, which rises from 0.615 to 0.635 as reported by Bulent *et al.,* (2006). This is similar to conclusions reached by De Figueiredo *et al* (2011) on sunflower seeds with different structural characteristics, Tunde and Akintunde (2004) on sesame seed from 0.54 to 0.60, Vilche *et al.,* (2003) on Quinoa seeds obtained increase from 0.77 to 0.80 with variation of moisture, and Abalone *et al.,* (2004) on Amaranth seeds. However, Gupta and Prakash (1992) did not find any trend between the sphericity of the seed and moisture content of JSF-1 safflower. On the other hand, Sedat *et al.,* (2005) on rapeseed (*Brassica napus oleifera L*), concluded that sphericity decreased with moisture from 0.93 to 0.91 within moisture range of 4.70 to 23.96%. Some researcher uses the trace of seed on a plain paper to calculate the sphericity of a seed. Other methods include the use of shadow or even photograph.

5.0 Surface area

Surface area is an important property of seed. It helps the designer in estimating the hopper, processing chamber and the chute. The surface area of some seeds can be obtained from the relation 5. In some seeds, the surface area has to be traced on the seed at natural resting position, and then the traced shape is enumerated in terms of surface area.

$$S = \pi D_g^2 \hspace{4cm} 5$$

Several researchers concluded increase in area of seeds with moisture, Konak *et al.,* (2002) for chick peas, Carman (1996) for lentil and Ozarslan (2002) for cotton seeds. For *Jatropha* seed, the surface area increases linearly from 476.78 to 521.99 mm^2 within moisture range of 4.75 to 19.57% as concluded Garnayak *et al.,* (2008) on dry-basis. These findings are similar to conclusions drawn by many researchers as reported in Olajide and Ade-Omowaye (1999) on locust bean seed, Aviara *et al* (1999) on guna seeds, Singh and Goswani (1998) on mechanical properties of cumin (*cuminum cyminum Linn)*, Faruk and Kubilay (2005) on pine *(Pinus pinea)* nuts and Panmanas *et al* (2007) on green soybean.

6.0   Porosity

Porosity is defined as the ratio of "empty" space of seed to its total volume. The porosity of seeds is expressed as a function of bulk and true densities (Mustafa, 2007). It is always between 0 – 1, expressed as a %. Equation 6 represents the relation. This is also dependent on the moisture content of the seed. For Jatropha oil seed, the porosity increase with the moisture content, but for neem nut, as reported by Visvanathan *et al.,* (1996) the porosity decreases. Mustafa (2007) reported on barbunia seeds that the porosity increases with moisture. He recorded an increase from 47.85- 48.56% with moisture content rising from 18.33 to 32.43%. Similar report was done by Konak *et al.,* (2002) on chick seeds, Kunlun *et al.,* (2009) on brown rice, Yalcin and Karababa (2007) reported on flaxseed. Karababa (2006) concluded same in popcorn kernels, as Majdi and Taha (2007) concluded on green wheat. The increase or decrease on the porosity is dependent on seed as seen from the conclusions of various researchers;

$$\varepsilon = \left(1 - \frac{\rho_b}{\rho_t}\right) x\ 100 \hspace{2cm} 6$$

ε is the porosity, ρ_b and ρ_t are the bulk and true densities respectively in kgm^{-3}.

7.0   Angle of repose

It is the angle made by seed, simply when heaped as a function of height of the heap and its base diameter. The angle of repose is determined by pouring the seeds in a container of known base diameter and height. When full, the container is slowly lifted until it is free of the seeds. The cone-shape made by the seeds is then measured in terms of diameter and height. The angle of repose is evaluated using equation 7. Sirisomboon *et al.,* (2007) and Ganayak *et al.,* (2008) reported on *Jatropha* to be in the range of 28 to 40^0. The variation of angle of repose with seed moisture is, as reported and concluded by many, that it increases linearly with the seed moisture content. Chandrasekan and Viswanathan (1999) on coffee, Jain and Bal (1997) on pearl millet, Suthar and Das (1996) on kirangda seed and many others. They are related as agreed by many with the friction between the seeds, since as the moisture level increases, the surface of the seeds become wet, thereby sticking to one another. From the review of several researchers' reports, no specific shape of material or size is adopted, but the range of 50 mm to 150 mm for base and 75 mm to 150 mm for height were predominantly used. This also depends of the sizes of the seed under investigation. Most of the seeds that are small, lower base diameters were adopted and height usually less than 100 mm. H is the height of the cone (cm) and D is the diameter of cone (cm).

$$\theta = \tan^{-1}\left(\frac{2H}{D}\right) \qquad 7$$

8.0 Static coefficient of friction

Coefficient of friction is the degree of resistance of seed to flow on a given surface. Various surfaces have different values of coefficient of friction. It is useful in design of processing and storage plants. It is evaluated as expressed in equation 8. The seed is carefully poured in a container of pre-determined diameter and height. It is full to brim and slowly lifted to ensure no contact between the container and the surface to which the coefficient of friction is to be determined. The surface is lifted gently until the container begins to slide down the surface. The vertical height at which sliding begins and the distance from base of the surface and the base of the vertical stand is measured as D. Usually, the D value for a specific surface is constant, but the value at which the sliding begins, h will vary with moisture content. The presence of water in seeds as the moisture content increases explains the cohesive force exerted by the seeds to the surface. As a result, the higher the moisture content, the higher the value of the coefficient of friction. This may increase with the test surface, as it will contribute to the friction. Various researchers have used different surfaces, such as metal sheet, aluminium sheet and plywood, and have concluded that the plywood offers more friction than others as concluded by Dutta *et al* (1988), Visvanathan *et al* (1996), Kulkelko *et al* (1998) and Shepherd and Bhardwaj (1986).

$$\mu = \tan \alpha \qquad 8$$

μ is the coefficient of friction, h is the height of cone and D is the base diameter of the cone.

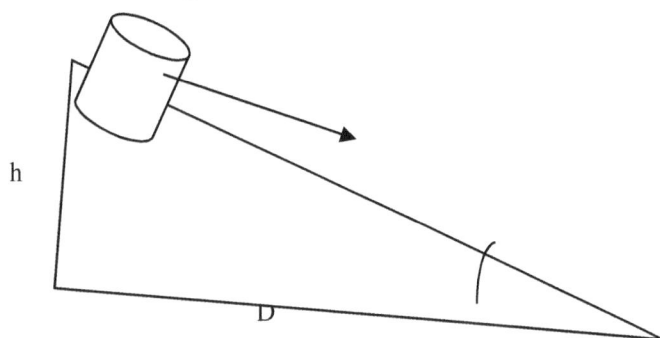

9.0 Seed mass

This is usually measured in 1000. The mass of 1000 seeds is used as a reference to determine the mass of a seed. This is used because most seeds are small and have negligible and varying weights. A calculated number of the seeds, pending the number chosen by the researcher, is individually counted and weighed. A good number of replicates to reduce error as low as possible is usually taken, say 10. Each replica is carefully weighted and the average of all readings is taken to be the seed mass. No specific equation is used in the determination of the seed mass by this method. The increase in the amount of water content in the seed increases the mass of the seed. Several researchers concluded increase in 1000 seed mass on various seeds with increase in moisture content. This is related to the size of a seed. However, where the seeds are big or of appreciable weight, 100 seed mass or 50 seed mass or even less are adopted. Sacilik *et al.,* (2003), Visvanathan *et al.,* (1996) and Ganayak *et al.,* (2008) concluded same on hemp seed, neem nut and jatropha respectively. Other researchers similarly found out that the mass of the seed is a function of moisture content. Konak *et al.,* (2002) on chick pea seeds, Makanjuola (1972) on some melon seeds, Sila *et al.,* (1993) on Tamarind (*Tamarindus indica*) seeds.

10.0 Bulk density

This is the density of "bulk" of seed as a function of a fixed volume of a container. This is done by freely pouring the seeds in a fixed volume container of predetermined weight and re-weighted to find the mass of seed. The density is then enumerated from the mass of seeds as a function of volume of the container. The seeds are parked in the container freely, usually dropped from a height of 15 cm, without compaction (Dutta et al., 1988; Garayak et al., 2008),. Increase in mass of the seed increases or decrease bulk density of the seed. Some researchers have presented negative trends on some seeds, while some reported otherwise. It is a function of seed. Shepherd and Bhardwaj (1986), Gupta and Das (1997) and Carman (1996), Kingsly et al., (2006) on dried pomegranate seeds (anardana), Oje and Ugbor (1991) on oil-bean seeds, Edward (2001) on bambara groundnut and Zewdu and Solomon (2007) on Tef seeds.

11.0 True density

This is the actual density of individual seed. It is a function of pre-known mass of seed to the volume of fluid displaced. In this case, a known mass of seed is placed in a known level of fluid. The volume of displaced liquid as a result of insertion of the seeds is measured and in relation to the mass, density is calculated. The fluids usually utilized for this are water, oil, toluene or gas (Dutta et al., (1988), Garayak et al., 2008; Konak et al., 2002). This has been concluded by Aviara et al., (2005) for Balanites aegyptiaca nuts, Konak et al (2002) for chick seeds, and Ganayak et al (2008) for jatropha, Onder et al (2007) on cowpea seeds (Vigna sinensis L), Juana et al., (2008) on Roselle seeds (Hibiscus sabdariffa, L) and Amin et al (2004) on lentil seeds.

12.0 Hardness of seeds

This is one of the engineering properties of seed. Seed is placed in either vertical or horizontal positions. A compression machine (like Instron machine) is used to find the force required to break the seed's hull. The feed is usually at a very low speed, ranging from 0.5 mm/min to 2 mm/min for most seeds. At the crack of the hull, the measured force to crack is recorded and plotted. For a mechanical system to be designed for processing seeds, amongst other properties, hardness of seed is important. For dehulling, shelling or husking, these factors affect the performance of mechanical system. The knowledge of the hardness of seed helps the designer in material selection. Some researchers have considered the surface and inner strength of seeds. Akaaimo and Raji (2006) on the prosopis Africana seeds presented that the crushing strength of the seeds was determined both by the pod and seed using California bearing ratio (CBR) machine. They concluded that a mean of 142.86 Nmm^{-2} was required, much more than the pods. This was similar to the works of Haque et al., (2000) on orange, Dinrifo and Faborode (1993), Shkelqim and Joachim (2010) on Jatropha seeds and kernels, Anonymous (1989) on pumpkin (Cucurbita moschata) seeds, Omobuwajo et al., (1999) on African breadfruit (treculia Africana) seed.

13.0 Conclusion

A comprehensive review of the methodologies for the determination of physical properties of seeds has been presented. The findings and conclusions of most of the researchers are similar and the methodology adopted for the determination of physical properties is similar. Conclusions drawn by several authors were studied and although using different seeds, most of the properties are determined to make the design of processing system possible. Moisture content has been concluded to be the main controlling factor in development of any seed processing device.

References

Abalone, R, Cassinera. A, Gaston. A, and Lara M.A. (2004). Some physical properties of Amaranth Seeds. *Biosystem Engineering, 89(1)*, 109-117

Akaaimo, D.I and Raji, A.O. (2006). Some physical and engineering properties of *Prosopis Africana* seeds. *Journal of Biosystems Engineering, 95(2)*, 197-205

Akobundu, E.N.T, Cherry, J.P and Simmons J.G. (1982) Chemical, functional and nutritional properties of egusi seed proteins products. *Food science, 47(3)*, 829-835

Akubuo, C.O and Odighoh E.U. (1999). Egusi fruit coring machine. *Journal of agricultural research, 74*, 121-126

Altuntases, E.O, Zgo, Z.E, Taser, O.F. (2005). Some physical properties of fenugreek seeds. *Journal of food Engineering, 71,* 37-43,

Amin, M.N, Hossain M.A and Roy, K.C. (2004). Effects of moisture content on some physical properties of lentil seeds. *Journal of food engineering, 65,* 83-87

Anonymous, (1989). Some Engineering properties of pumpkin (*cucurbita moschata*) seeds. *Journal of Food Engineering, 9,* 153-162

Avaira, M.E, Umar, B. (2005). Some physical properties of *Balanites aegyptiaca* nuts. *Biosystems Engineering, 92(3),* 325-334,

Aydin, C, Ogut, H and Konak M. (2002). Some physical properties of Turkish Mahaleb. *Biosystems Engineering, 82(2),* 231-234

Azmi, Y. and Stephen, J.M. (1994). Physical and chemical characterization of methyl soyoil and methyl tallow esters as CI engine fuels. *Biomass and Bioenergy, 6(4),* 321-328

Babasaheb, B.D. (2004). Seeds Handbook, Biology, Production, Processing and Storage, Second edition, Marcel Dekker Inc, NY, 1-4,

Bal, S and Mishra, H.S. (1988). Engineering properties of soybeans. *Proceedings of the National seminar on soybean processing and utilization in India,* 146-165

Bulent, M.C, Ibrahim, Y, Cengiz, O. (2006). Physical properties of sweet corn seed (Zea mays Sacc). *Journal of Food Engineering, 74,* 523-528

Carman, K. (1996). Some physical properties of lentil seeds. *Journal of Agricultural Engineering Research, 63,* 87-92

Chandrasekan, V and Viswanathan, R. (1999). Physical and thermal properties of coffee. *Journal of agricultural engineering Research, 73*, 227-234

De-figueiredo, A.K, Baumler, E. Riccobene, I.C and Nolasco, S.M. (2011). Moisture-dependent engineering properties of sunflower seeds with different structural characteristics. *Journal of food engineering, 102*, 58-65

Dinrifo, R.R and Faborode,M.O. (1993). Application of hertz theory of contact stresses to cocoa pod deformation. *J. of Agric Engg and Technology, 1*, 63-73

Dursun, E. and Dursun I. (2005). Some physical properties of caper seed. *Journal of Biosystem Engineering, 95(2)*, 237-245

Dursun, I, Tugrul, K.M and Dursum, E. (2007). Some physical properties of sugarbeet seed. *Journal of stored products Research, 43*, 149-155

Dutta, S.K, Nema, V.K and Bhardwaj, R.K. (1988). Physical properties of gram. *Journal of Agricultural Engineering Resources, 39*, 259-268

Ebubekir. A, Engin, O.O and Faruk, T. (2005). Some physical properties of fenugreek (*trigonella foenum-graceum L*) seeds. *Journal of Food Engineering, 71*, 37-43

Edward, A.B. (2001). Physical properties of bambara groundnuts. *J. of Food Engineering, 47*, 321-326

Erica, B, Adela C, Susanam M.N and Isabel C.R. (2006). Moisture dependent physical and compressive properties of safflower seeds. *J. of Food Engg, 72*, 134-140

Garnayak, D.K, Pradhan, R.C, Naik,S.N and Bhatnagar, N. (2008). Moisture-dependent physical properties of Jatropha seed (jatropha curcas L). *Journal of Industrial crops and products, 27*, 123-129

Gupta, R.K. Das, S.K. (1997). Physical properties of sunflower seeds. *Journal of Agricultural Engineering Research, 66(1)*, 1-8

Gupta, R.K and Prakash, S. (1993). The effect of seed moisture content on the physical properties of JSF-1 safflower. *J. of oilseeds Research 9(2)*, 209-216

Haque, M.A, Aviara N.A and Mamman, E, (2000). Development and performance evaluation of domestic orange juice extraction. *Journal of Agricultural Technology, 8(2)*, 32-38

Jain, R.K and Bal, S (1997). Properties of pearl millet. *Journal of Agricultural Engineering Research, 66* 85-91

Juana, S.M, Aurelio, D.L, Salvado, N.G and Jose A.L.S. (2008). Some physical properties of rosell (*hibiscus sabdariffa L*) seeds as a function of moisture content. *Journal of Food Engineering, 87*, 391-397

Kingsly, A.R.P, Singh, D.B, Manikantan, M.R and Jain, R.K. (2006). Moisture dependent physical properties of dried pomegranate seeds (*anardana). Journal of Food Engineering, 75*, 492-496

Konak, M. Carman, K and Aydin, C. (2002). Physical properties of chick pea seed. *Biosystem Engineering, 82(1)*, 73-78

Kulkelko, D.A, Jayas, D.S, White, N.D.G and Britton, M.G. (1988). Physical properties of canola meal. Can. Agricultural Engineering, 30(1), 61-64

Makanjuola, G.A. (1972). A study of the physical properties of Melon seeds. *Journal of Agricultural Engineering Research 17,* 128-137

Sacilik, K. Ozturk, R and Keskin, R. (2003). Some physical properties of hemp seed. *Biosystems Engineering, 86(4),* 441-448,

Sahoo, P.K and Srivastava, A.P. (2002). Physical properties of okra seeds. *Biosystem Engineering, 83(4),* 441-448

Sedat, C, Tamer, M, Huseyin, O and Ozden, O. (2005). Physical properties of rapeseed (*Brassica napus oleifea L).* *Journal of Food Engineering, 69,* 61-66

Sharma, V,Das, L. Prandhan, R.C, Naik, N.S, Bhatnagar, N and Kurell, R.S. (2011). Physical properties of tung seed: An industrial oil yielding crop. *Journal of industrial crops and products, 33,* 440-444

Shepherd,H and Bhardwaj, R.K. (1986). Moisture-dependent physical properties of pigeon pea. *Journal of Agricultural Engineering Research, 35,* 227-234

Shkelqim, K and Joachim M. (2010). Determination of physical, mechanical and chemical properties of seeds and kernels of (*Jatropha curcas L).* *Journal of industrial crops and products, 32,* 129-138

Sila B, Bal.S and Mukherjee, R.K. (1993). Some physical and engineering properties of Tamarind (*Tamaridus indica)* seed. *Journal of Food Engineering 18,* 77-89

Singh, K.K and Goswami, T.K. Physical properties of cumin seed. Journal of Agricultural Engineering Research, 35(4), 227-234, 1986

Sirisomboon, P. Kitchaiya, P. Pholpho, T and mahuttanyavanitch, W. Physical and Mechanical properties of Jatropha Curcas L fruits, nuts and kernes. Journal of Biosystems Engineering, 97(2), 201-207, 2007

Solomon, G, Luqman C.A and Mariah, A. (2010). Investigating egusi seed oil as potential biodiesel feedstock. *Energies, 3,* 601-618

Solomon, W.K and Zewdu, A.D. (2009). Moisture-dependent physical properties of niger seed (*Guizotia abyssinica Cass*). *J. of Ind.crops and prods, 29,* 165-170

Suthan, S.H and Das, S.K (1996). Some Physical properties of karingda seeds. *Journal of Agricultural Engineering Research, 65,* 15-22

Mohsenin, N.N. Physical properties of plant and animal materials. New York, Gordon and Breach Science Publishers, 1970

Mustafa, C. Physical properties of barbunia bean seed. Journal of Food Engineering, 80, 353-358, 2007

Ogunjimi, L.A.O, Aviara, N.A and Aregbesola, O.A (2002). Some Engineering properties of bean seed. *Journal of Food Engineering, 55,* 95-99

Ogut H (1998). Some physical properties of white lupin. *Journal of Agricultural Engineering Research, 86(2),* 191-198

Oje, K and Ugbor, E.C. (1991). Some physical properties of Oilbean seed. *Journal of Agricultural Research, 50,* 305-313

Oluba, O.M, Ogunlowo, Y.R, Ojieh, G.C and Adebisi, K.E. Physicochemical properties and fatty acid composition of seed oil. *J. of bio. sciences, 8(4),* 814-817, 2008

Omobuwajo, T.O, Akande, E.A and Sanni, L.A. (1999). Selected physical, mechanical and aerodynamic properties of African breadfruit (*treculia africana*) seeds. *Journal of Food Engineering, 40,* 241-244

Omobuwajo, T.O. Omobujo, R.O and Sanni, L.A. (2003). Physical properties of calabash nutmeg (*monodora myristica)* seeds. *J.of Food Engg, 57,* 375-381

Omobuwajo, T.O, Sanni, L.A and Balami, Y.A. (2000). Physical properties of sorrel (*Hibiscus sabdariffa)* seeds. *Journal of Food Engineering, 45,* 37-41

Onder,K. Erdem, Y, Aziz, O and Ibrahim A. (2007). Some physical and nutritional properties of cowpea seed (*vigna sinensis L*). *J.of Food Engg, 79,* 1405-1409

Oyelade, O.J, Odugbenro, P.O, Abioye, A.O and Raji, N.L. (2005). Some physical properties of African star apple (*chrysophyllus alibidum)* seeds. *Journal of Food Engineering, 67,* 435-440

Ozarslan, C. (2002). Physical properties of cotton seeds. *Biosystems Engi, 83(2),* 169-172

Paksoy, M and Aydin, C. (2004). Some physical properties of edible squash (*Cucurbita pepo L)* seeds. *Journal of Food Engineering, 65,* 225-231

Tunde A.T.Y and Akintunde, B.O. (2004). Some physical properties of sesame seed. *Biosystem Engineering, 88(1),* 127-129

Vanesa, Y.I, Susana, M.N and Mabel, C.T (2008). Physical properties of chia (*salvia hispanica L*) seed. *Journal of Industrial crop and products, 28,* 286-293

Vilche, C. Gely, M and Santalla, E. (2003). Physical properties of Quinoa seeds. *Biosystem Engineering 86(1),* 59-65

Visvanathan, R. Palanisamy, P.T, Gothandapani, L and Sreenarayanan V.V. (1996). Physical properties of neem nuts. *J. of Agric. Engi.g Research, 63,* 19-26

Yelcin, C. and Ersan, K (2007). Physical properties of coriander seeds (*coriandrum sativum L.)*. *Journal of Food Engineering, 80,* 408-416

Yelcin I, and Ozarslam C (2004). Physical properties of vetch seeds. *Biosystem Engineering, 88(4),* 507-512

Yelcin I, Ozarslan, C and Akbas, T (2007). Physical Properties of Pea (*pisum sativum)* seed. *Journal of Food Engineering, 79,* 731-735

Zewdu, A.D and Solomon W.K. (2007). Moisture-dependent physical properties of Tef seed. *Journal of Biosystem Engineering, 96(1),* 57-63

Chapter 4: Comparison of suppression subtractive hybridization with other methods used to identify differentially expressed genes in plants

M. Sahebi*,M.M.Hanafi,N. M Adam and P. Azizi
Laboratory of Plantation Crops,
Institute of Tropical Agriculture,
Universiti Putra Malaysia, Serdang, Selangor.
E-mail: mahbod_sahebi@yahoo.com

Abstract

In order to identify differential expressed genes (Target cDNA fragments), several techniques have been used such as suppression subtractive hybridization (SSH), Amplified Fragment Length Polymorphism PCR (**AFLP**-PCR), Differential display (DDRT-PCR or DD-PCR), Microarray and Gene Expression (MAGE). The SSH method numerously has been used to separate two different DNA molecules of target genes (tester) and strains of driver involving undesirable sequences (Akopyants et al., 1998; Diatchenko et al., 1996; Gurskaya et al., 1996; Luk'ianov et al., 1994). Suppression subtractive hybridization method works on suppression polymerase chain reaction (Lukyanov 1995; Siebert et al., 1995), and that is a single method include normalization and also subtraction during hybridization (Gurskaya et al., 1996).

Keywords: Target cDNA fragment, Driver, SSH, DDRT, cDNA- AFLP, Microarray

Introduction

The Process of gene expression demonstrates the information of gene employed to obtain functional product of gene,which including proteins, and non-protein products such as small nuclear RNA, ribosomal RNA and transfer RNA made by non-protein coding gene. The process of gene expression involves several steps: 1) Transcription 2) RNA splicing 3)

Translation and Post-translational reformation of a protein (Altschul at al., 1997). Cells obtain their abilities of controlling over function and also structure by gene regulation for creating cellular variation, adaptability and morphogenesis of any creature. The principal of gene expression is that genetic codes involved in DNA are interpreted and phenotype of organisms existed by expression attributes.

The profile of gene expression in biology studies demonstrates a comprehensive cellular function and is able to differentiate cells with different activities and reflects cell reaction to specific treatment. Since every gene belongs to a certain cell in terms of presenting, a wide range of experiments are needed to measure the whole genome of plants simultaneously. When genome sequencing is done, a confident step showing exactly what the cell doing at the moment, is expression profiling, whereas sequence of genome is just able to show what probably cell could do. Messenger RNA made by only a division of the genes, Presence or absence of mRNA in cells made by a gene at each moment indicatingthat the geneison or off respectively. Some elements carry the duty of determining whether a gene is off or on, including time of expression, cells in terms of actively dividing, also signals existed from adjacentcells. Different cells of different parts of plants are vary with each other by expressing slightly altered genes. Hence an expression profile provides a situation decreasing cells type and location. An experiment of expression profiling quantifies the comparative volume of mRNA expressed in different experimental settings, since different levels of a particular sequence of messenger RNA offer changing in coding of proteins.

Suppression subtractive hybridization

Suppression subtractive hybridization (SSH) method acts as a beneficial instrument to identify differentially regulated genes valuable for cells in terms of growing and being differences (Lukyanov at al., 2007).During subtraction the large quantity of DNA fragments become equal by normalization (Diatchenko et al., 1996; Gurskaya et al., 1996; Jin, X, L, PD,

& CC, 1997)meaning that the concentration of high and low abundant cDNA fragments become equal, although practically it is not possible to equal all the differentially expressed genes. The amount of one specific cDNA fragments depends on some factors that involve the initial value of cDNA fragments, the subtracted proportion of cDNA concentration in the sample, and the other genes differentially expressed. Success rate of suppression subtractive hybridization significantly depends on initial material, the proportion of two samples (tester and driver) that are being subtracted(Lukyanov at al., 2007). Suppression subtractive hybridization (SSH) is quite operative almost in many cases though while samples under investigation show a few number of differences sequences, and subtracted libraries obtained from SSH to be involved a great deal of background with Mirror Orientation Selection (MOS) pointedly diminish the high back ground. Consequently there are many methods that have been used during recent decade to isolate important genes (Hara, T, S, S, & K, 1991; Hedrick, DI, EA, & MM, 1984; Hubank, DG, & 1994; Sargent & IB, 1983; Wang & DD, 1991). But, their advantages are reduced because of complication of the methods and demand of a great deal of starting material (Total RNA). There for finding a method with requirement of minimum of initiative material will be greatly useful and more efficient. (Lukyanov at al., 2007).

False positive obtained from different clones which create distinct signal during initiative screening process but not able to be confirmed by following analysis of details, is omitted by employing MOS system. Mirror Orientation Selection significantly reduces the amount of primary clones (Rebrikov et al., 2000; Rebrikov, SM, PD, SA, & 2004).

The rationale of Mirror Oriented Selection or (MOS) is that during the SSH method while PCR amplification be done, each type of molecule existing in background just involve one orientation corresponding to inverted terminal repeats (adapter)(Rebrikov et al., 2004), that is related to the orientation of primitive molecule. In contrast, because of effective enrichment during PCR amplification, the interested cDNA fragment exists there, hence each particular sequence is inclusive lots of progenitors and will represented by both relative orientations to inverted terminal repeats (adapters) (Rebrikov et al., 2000; Rebrikov et al., 2004).

The principle of suppression subtractive hybridization is that target cDNA fragments are amplified selectively whereas non-target cDNA are suppressed from amplification during subtractive hybridization when long inverted terminal repeats ligated to 5´ends of single strand cDNA furthermore prevent from their amplification. Generally suppression subtractive hybridization involve four main steps including: construction cDNA library which involves isolation total RNA from both types of cell population (Tester and Driver) following that the synthesized double- stranded cDNA. During second step ds-cDNA from previous step be digested by restriction enzymes which pluck the ends, like RsaI or AluI (Rebrikov et al., 2004). The third step involves preparation two different inverted terminals (adapters) needed for hybridization and finally hybridization step (Lukyanov at al., 2007).

Amplified Fragment-Length Polymorphism

Amplified fragment-length polymorphisms(ALFP) using other method which widely is employed to genes analyses of specific biological processes. This method significantly has been employed to discover different genes (Biezen, H, JE, & JDG, 2000; Durrant, O, P, KE, & JDG, 2000; Kornmann, N, D, F, & U, 2001; Qin et al., 2000; Sutcliffe et al., 2000).But generally it is not efficient for quantitative analysis of gene expression(Breyne et al., 2003a). In order to prepare classification fragments of AFLP, after cutting bands related to transcription of differentially expressed genes, eluting the DNA, and then reamplified with the same qualification utilized for optional amplification, following that PCR product obtained from reamplification, sequenced by the elective primer. The next step is employing pGEM-T easy vector to clone the fragments, and then comparing newly sequence by BLAST database (Altschul et al., 1997). The cDNA –AFLP method is not dependent of sequence data, hence it is useable for differentially expressed genes analysis and also widely used for biological studies, expressly while lack of genomic resources are still existed. By utilizing microarray technology simultaneously, cDNA –AFLP will be used for gene discovery and also for comprehensive gene expression studies (Breyne et al., 2003b). The most limitation of DNA –AFLP methods among other method is the existing heterogeneity observed in the final products.

Analysis of microarray gene expression data

The other widely useable method for identification and functional studies of comprehensiveness of the genome is the microarray expression analysis. Microarray analysis provides situation of determining

relative levels of messenger RNA is great quantity in just one set of cell population (tissue) for thousands of genes at a time, where microarray works on a series of computational analysis techniques.(Huber, Heydebreck, & Vingron, 2003). The DNA microarray establishes an outstanding pattern. The purpose of this technology is to measure the levels of mRNA in specific tissue or cell simultaneously. The ss-cDNA of our target gene (interest genes) –which might be so in thousands- will be stopped by presence on the slide of glass, a wafer of quartz, or a membrane of nylon to arrange in a mesh (array). After extraction of total RNA, following that extraction of messenger RNA, each sequence will be labeled and then hybridized for array. Quantity and yield value of label must correspond to the amount of transcription of RNA in the sample (Huber et al., 2003).

Besides the most advantagesof thismethod providing the ability to monitor the function of one particular gene, and also its interaction with the other genes in comprehensive genome on a solitary chip. There are some disadvantages utilizing microarray techniques as well, for instance we can refer to: The short percentage of survived gene which either a clone of cDNA or a sequence of DNA; hardness of separating among distinct transcription of genes to unit family; high sensitivity to hybridization consequently requirement a great amount of RNA.

Differential-Display Reverse Transcription

The other technique used by researchers which makes them able to determine and compare changes in gene expression at the level of mRNA among any pair of cell samples of eukaryotic, is differentially displayed or (DDRT-PCR). Two samples are different from each other morphologically and also genetically so scientists are looking for the patterns of gene expression, via finding the base cause of the specific or difference genes affected by experimentation. The principle of this method is using a few amount of short random primers combining with anchored primers or oligo-dT in order to amplify and imagine the majority of messenger in a particular cell. In order to having better result with lesser false-positive through DDRT-PCR method fine –quality of DNase without any template of RNA is needed, also RNA always should be treated by DNase(Liang et al., 1993). False-positive among the results seems to be the significant restriction of Differential-Display Reverse Transcription technique caused by laxity annealing process, isolation products of PCR with impurity, and using only desirable primers for amplification, so some steps are needed to overcome these problems involving modification through design of primer and re-amplification (Sturtevant, 2000).

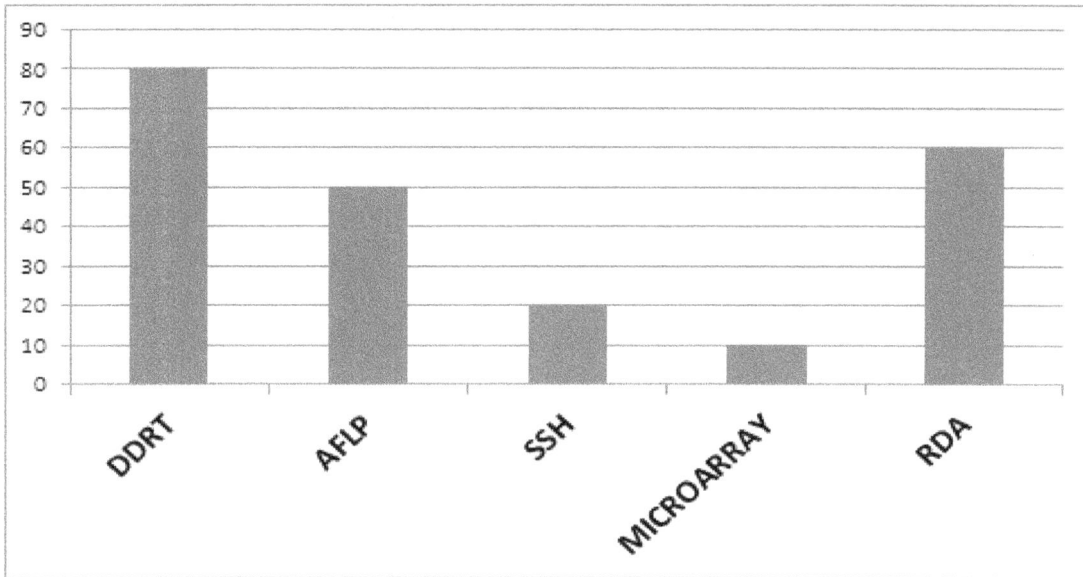

Figure1. RATE OF FALSE POSSITIVE AMONG DIFFEREN TMETHODS

Conclusion

For false positive appears among the results, the SSH technique is much better than cDNA-AFLP, RDA and DDRT-PCR and in terms ofcosteffectiveness the SSH method is more affordable than the microarray technique. Although the false positive of microarray technique is also fewer like the SSH method or much fewer but it's more costly (Figure1, 2 and Table 1).

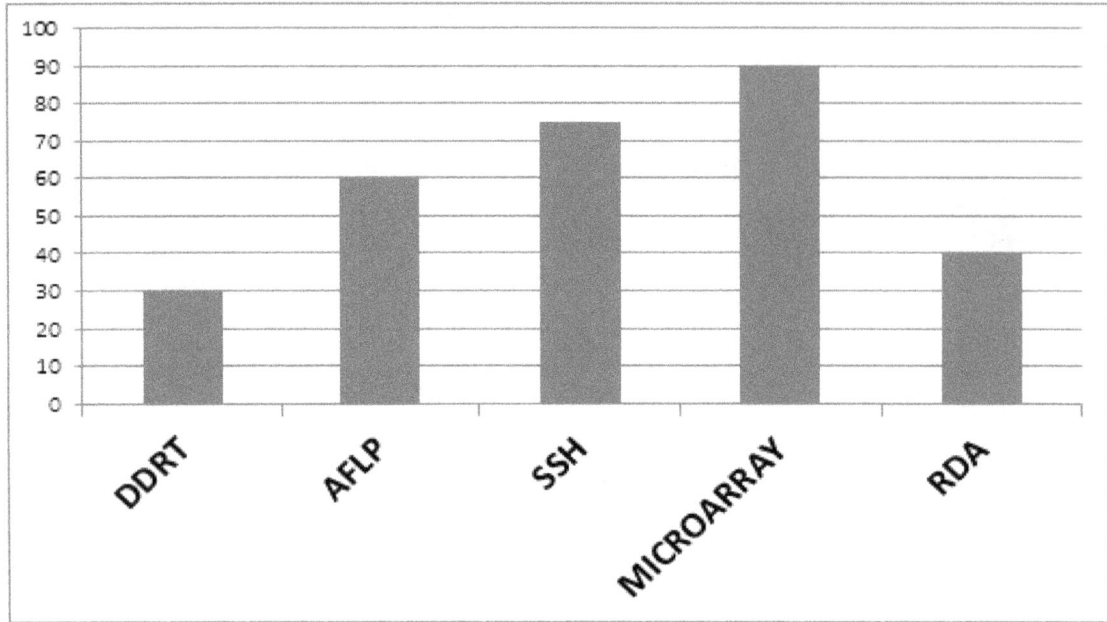

Figure2. SUCCESSFUL RATE OF DIFFERENT METHODS

Table 1: Comparison of the test characteristics

	DDRT	AFLP	SSH	MIROARRAY	RDA
Amount of required RNA	H	H	M	H	M
Complication	M	H	H	H	L
Coast	M	M	M	H	L
False positive	H	H	L	L	H
H: High M: Medium L: Low					

References:

Akopyants. (1998). PCR-based subtractive hybridization and differences in gene content among strains of Helicobacter pylori. *Proc Natl Academic Science USA, 95,* 13108-13113.

Altschul, s. (1997). Gapped BLAST and PSI-BLAST: a new generation of protein database search programs. *Nucleic Acids Res, 25,* 3389-3402.

Biezen. (2000). cDNA-AFLP display for the isolation of Peronospora parasitica genes expressed during infection in Arabidopsis thaliana. *Mol Plant-Microbe Interact, 13,* 895-898.

Breyne, P., Dreesen, R., Cannoot, B., Rombaut, D., Vandepoele, K., Rombauts, S. (2003a). Quantitative cDNA-AFLP analysis for genome-wide expression studies. *Mol Gen Genomics, 269,* 173-179.

Breyne, P., Dreesen, R., Cannoot, B., Rombaut, D., Vandepoele, K., Rombauts, S. (2003b). Quantitative cDNA-AFLP analysis for genome-wide expression studies (Publication., from Mol Gen Genomics:

Diatchen. (1996). Suppression subtractive hybridization: a method for generating differentially regulated or tissue-specific cDNA probes and libraries. *Proc Natl Academic Scienc USA, 93,* 6025-6030.

Durrant. (2000). cDNA-AFLP reveals a striking overlap in race-specific resistance and wound response gene expression profiles. *Plant Cell, 12,* 963-977.

Gurska. (1996). Equalizing cDNA subtraction based on selective suppression of polymerase chain reaction: cloning of Jurkat cell transcripts induced by phytohemaglutinin and phorbol 12-myristate 13-acetate. *Anal Biochem, 240,* 90-97.

Hara, E. (1991). Subtractive cDNA cloning using oligo(dT)30-latex and PCR: isolation of cDNA clones specific to undifferentiated human embryonal carcinoma cells. *Nucleic Acids Res, 19,* 7097–7104.

Hedrick, S. (1984). Isolation of cDNA clones encoding T cellspecific membrane-associated proteins. *Nature, 308,* 149-153.

Hubank, M., DG, S., & (1994). Identifying differences in mRNA expression by representational difference analysis of cDNA. *Nucleic Acids Res, 22,* 5640-5648.

Huber, W., Heydebreck, A. v., & Vingron, M. (2003). *Analysis of microarray gene expression data.*Unpublished manuscript.

Jin. (1997). Differential screening of a subtracted cDNA library: a method to search for genes preferentially expressed in multiple tissues.

. *Biotechniques, 23,* 1084-1086.

Kornmann, B. (2001). Analysis of circadian liver gene expression by ADDER, a highly sensitive method for the display of differentially expressed mRNAs. *Nucleic Acids Res, 29,* e51.

Luk'ianov, S. (1994). Highly-effective subtractive hybridization of cDNA. *Bioorg Khim, 20,* 701-704.

Lukyanov, K. (1995). Inverted terminal repeats permit the average length of amplified DNA fragments to be regulated during preparation

of cDNA libraries by polymerase chain reaction. *Anal Biochem, 229,* 198-202

P Liang, Averboukh, L., Pardee, A. B., & (1993). Distribution and cloning of eukaryotic mRNAs by means of differential display: refinements and optimization. *Nucleic Acids Res, 21,* 3269-3275.

Qin, L. (2000). An efficient cDNA-AFLP-based strategy for the identification of putative pathogenicity factors from the potato cyst nematode Globodera rostochiensis. *Mol Plant-Microbe Interact, 13*, 830-836.

Rebrikov, D. (2000). Mirror orientation selection (MOS): a method for eliminating false positive clones from libraries

generated by suppression subtractive hybridization. *Nucleic Acids Res, 28*, e90.

Rebrikov, D.(2004). Suppression subtractive hybridization. *Methods Mol Biol 258*, 107-134.

Sargent, T., & IB, D. (1983). Differential gene expression in the gastrula of Xenopus laevis. *Science, 222*, 135-139.

Siebert, P. (1995). An improved PCR method for walking in uncloned genomic DNA. *Nucleic Acids Res, 23*, 1087-1088.

Sturtevant, J. (2000). Applications of Differential-Display Reverse Transcription-PCR to Molecular Pathogenesis and Medical Mycology. *Clinical microbiology reviews, 13*(3), 408.

Sutcliffe, G. (2000). TOGA: an automated parsing technology for analyzing expression of nearly all genes. *Proc Natl Acad Sci USA*

97, 1976-1981.

Wang, Z., & DD, B. (1991). A gene expression screen. *Proc Natl Acad Sci USA, 88*, 11505-11509.

Chapter 5: Impact of Blast Disease on Rice *(Oryza sativa L.)*

P. Azizi, *M.Y. Rafii, N.M. Adam, and M. Sahebi

Laboratory of Plantation Crops,
Institute of Tropical Agriculture,
Universiti Putra Malaysia, Serdang, Selangor.
E-mail: bahar3236@yahoo.com

Abstract

Rice is one of the most important cereals and a food resource in the world. Over half of the world population has depended to the rice as a main food. The Ascomycete fungus, *Magnaporthegrisea* Barr (anamorph*Pyriculariagrisea*Sacc., synonym *P.oryzae*Cav.) is source of blast disease, one of the dangerous significant diseases of rice, that is disturbing rice. The pathogen can pass on a disease to all the portions of a rice plant at different growing steps: leaf, collar, internode, node, or neck, and other parts of the panicle, and from time to time the leaf sheath. So, Blast disease annually is decreased and destroyed yield of rice. There are different biological and chemical techniques for the control of rice blast, which are complex using of healthy seed, controlling of fertilizer, cultural system, chemical management, and consuming of resistant cultivar.

Keyword: blast disease, pathogen, fungus, resistant cultivar.

Rice *(Oryza sativa L.)*

After wheat, rice *(Oryza Sativa L.)* is the second main produced cereal in the world. As a cereal grain, it is the greatest significant staple food for a huge portion of the world's people (Couch & Kohn, 2002). It has two main subspecies involve indica and *japonica*. In addition the genus Oryza encompasses of 21 wild families of the farm rice. There are four species O. sativa, O. officialis, O. ridelyi and O. granulatein Oryza genus. All of the Oryza genus of rice have n = 12 chromosomes and while interspecific crossing is possible inside each multipart, it is hard to improve fertile progeny from crosses through multiplexes(Vaughan, Morishima, & Kadowaki, 2003). The O. sativa combined covers two farm species: O. sativa and O. glaberrima, and five or six wildspecies: O. rufipogon, O. nivara (also measured to be anecotype of O. rufipogon), O. barthii, O. longistaminata,O. meridionalis and O. glumaepatula, all of them arediploids. Oryza sativa is spread totally with a high focus in Asia and was domesticated from O. rufipogon, while O. glaberrima is mature in West Africa (Bres-Patry, Lorieux, Clement, Bangraz, & Ghesquiere, 2001; Cai & Morishima, 2002; Li, Zhou, & Sang, 2006; Thomson et al., 2003; Uga et al., 2003; Xiao et al., 1998; Xiong, Liu, Dai, Xu, & Zhang, 1999).

Importance of the blast disease

Rice blast is the maximum distractive disease regarding the rice crop in the world. Ever since rice is a main food resource for more than 50% of the world population, its sound effects have a broad range. It has been originated in above 85 countries through the world. Yearly the quantity of crops missing to rice blast could nourish 60 million of general public. Even though there are some resistant strains of rice, the disease keeps it up everywhere rice is grown up. The disease has never been damaged from an area. It has injured plants by production of spores and penetration of infection (Hajime, 2001). Because of having negative effects and dropping of yield of rice we prepared this article review in order to give some more information about it and measure the influence of blast disease on rice.

Causative organism

The Ascomycete fungus, *Magnaporthegrisea* Barr (anamorph*Pyriculariagrisea*Sacc., synonym *P. oryzae*Cav.) is provided Rice blast. They are about 20×10 µm, shaped on conidiophores which overhang from lesions on plants. These incubate and develop an appressorium, (Figure 1) at the gradient of the germ tube, which attaches tothe external of plant tissues; an infection-peg from the appresorium penetrates into plant tissues. The wall of conidiophores and appressorium are pigmented by melanin(Hajime, 2001).

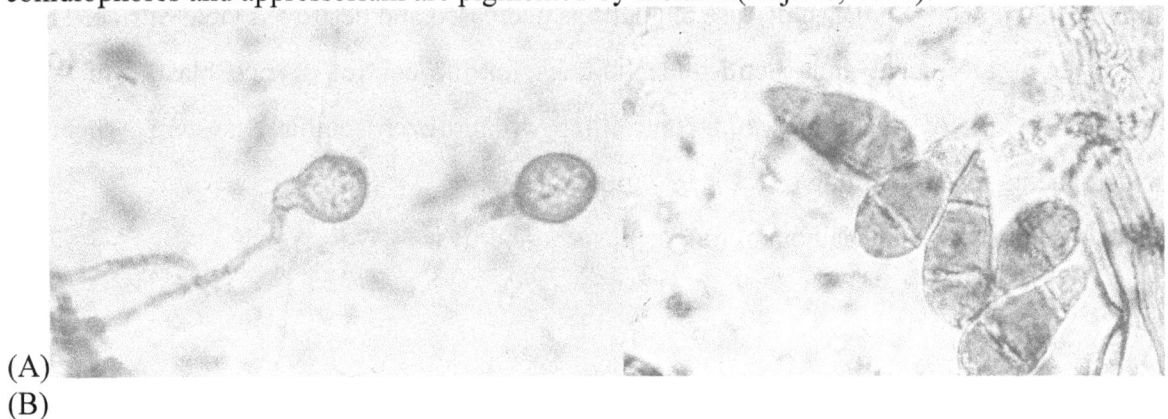

(A)
(B)

Figure1: Appressoria(A), Macroconidia(B)

Warning sign

Blast has injured plants by production of spores and penetration of infection.

Leaf Signs

Lesions are taking place the leaf, usuallydiamond-shaped and are 0.39 to 0.58 inch (1.0-1.5 cm) long and 0.12 - 0.2 inch (0.3-0.5 cm) extensive. The colors of them are gray or white medium, brown or reddish brown edge, white or grey-green middle and a darker green edge. These characters of lesionsare depended on varietal resistance, age of the plant, and lesion age. Sometimes leaf blast has died the young plants in the tiller stage completely. This disease usually increases early in the season then declines late in the season as leaves become less susceptible.

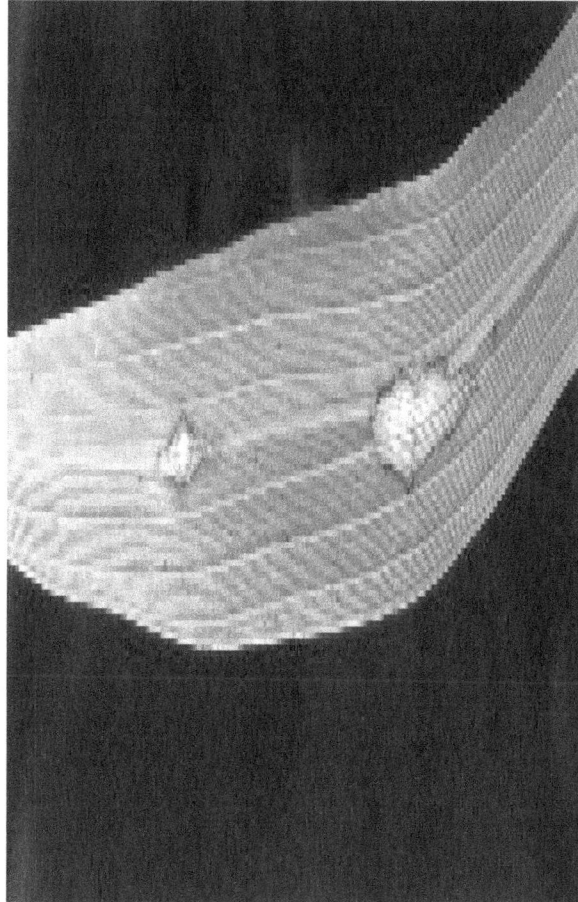

Figure 2. Blast lesions on the leaf

Neck decay and panicle blast

Reaction of rice necks or panicles to blast have usually been measured to be parallel to those of leaves (Ou, 1985). Node and nearby zone at base of panicle discolored brown or chocolate brown ; stem of panicle dry up and may perhaps breakdown; node purplish or blue-gray with conidia of the fungal pathogen; panicle white or gray; florets do not block up and fit gray; panicle branches and stems of florets with gray-brown lesions. Pollution to the neck node creates wedge-shaped purplish lesions, followed by lesion elongation to both sides of the neck node – signs which are severe for grain improvement. While young neck nodes are attacked, the panicles turn into white in color – the so-called 'white head' that is occasionally misunderstood as insect injury. Later pollution reasons imperfect grain filling, and reduced grain value. Panicle branches and glumes can be infested too. Spikelet's attacked by the fungus modification to white in color from the upper and create several conidia, which develop the inoculum foundation after heading.

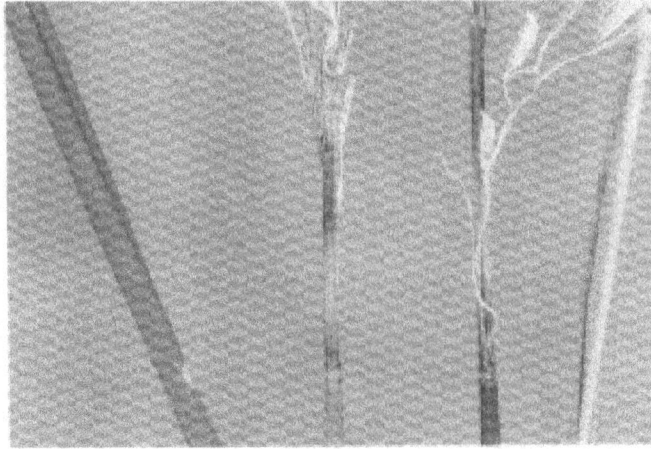

Figure3.Panicles, Florets, and Grain, rotten neck blast

Collar rot

The collar rot stage happens in arrears to infection at the intersection of the leaf blade and sheath, ensuing in growth of appearances brown to dark brown lesion.Severe collar rot infection on the flag or second to last leaf often kills the whole of leaf.

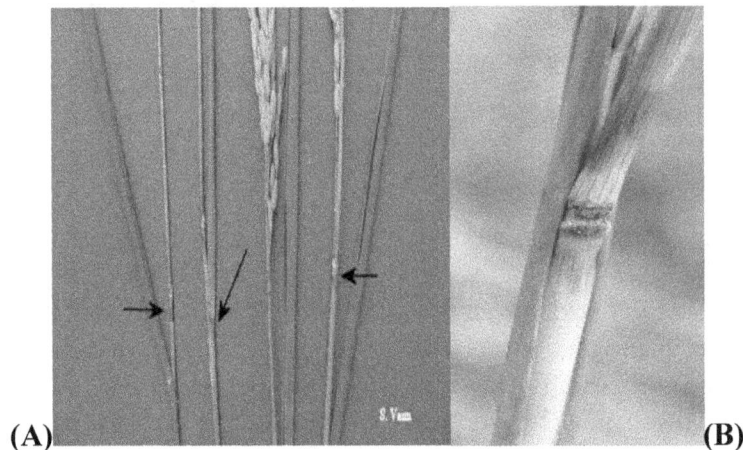

(A)   (B)

Figure4.(A) Collar rot, (B) Node blast

Node blast

The nodes at the base of the panicle and the branches of panicle have taken on by blast fungus. A few amount of seeds will be exist inside the panicle consequently the head of panicle will turn lighter, usually called blast of rice. Also the node of panicle infected immorally, the panicle will be collapsed by illness of neck known the**"rotten neck"**.The through of panicle possibly will pass on if neck rot happens early. While a node is infested, the sheath tissue rots and the portion of the stem over head the point of pollution regularly are destroyed. In some cases, the node is in capacitated to the degree that the stems will disruption producing wide lodging.

Disease cycle

Generally, the epidemic of leaf blast disease is under control by two main factors infection (a vigorous leaf position is infected by pathogen spore) and sporulation (the number of spores created by a blast lesion in excess of an infective period), which are significant affective stages in control disease. The first infections on new seedling Tissues in the sleep through winter stage have been happened by the spores on leaves germinate and leaf tissues which are attacked. The amounts of tissues that are infested usually have connected with disease harshness. The lesions on the second step of infection are created more and more spores on the strong leaf materials, in addition these step could be reiterated several time through the developing period.Also different environmental factors, involving the amount of the water in the paddy, the using of nitrogen as fertilize and the fall of rain, even the temperature and genetically level of resistance of plants are affected on the number of cycles and spores.

Control measures

Vigorous seed

To take strong seeds, the seeds need be collected from the field situated below negative situations for the pathogen, and fungicide has to beused if needed. Gravity separation techniques for kernels also are beneficial.

Fertilizer organization

The method recommending is that using of correctly fertilize for every one of the altered cultivars. Nitrogen and phosphorus content in the plants moves up the Possibility of the disease. Additional nitrogen fertilizer is a cause of increasing disease, while using of silica decreases disease growth. So the amount and type of fertilizer have to be sensibly decided upon conferring to the cultivar are used, soil situation, climatic settings and disease possibility (Castilla, Savary, Cruz, & Hmarch, 2010).

Cultural methods

Planting into water reduces disease transition from kernels to seedlings as a result of the anaerobic situation that is disapproving to the pathogen. On the other hand, planting on damp soil promotes seed transition. Darkness moves disease happening as a consequence of the stretched wet circumstance. The important method that either effective yet simple is crop rotation. The reason for recommending it is that sustainable spores become apart in cropremains from the young seedling.

Chemical control

In order to manage blast disease several methods have been used such as using of fungicides which are the second fungicide market universal, also they have used in Broad vast of ricefields with the purpose of control blast. The populations of the pathogen will be resistant to the pathogen if we use fungicides with parallel types of act over extensive periods. so, it is not recommended (Kim, Oh, Hwang, & Kim, 2008; Long, Lee, & TeBeest, 2000).

Resistant cultivars

Resistant cultivars are, certainly, the main control choice for blast, regardless of the problems blast symbolizes in increasing durable and effective resistances. Blast resistance has been widely measured by molecular genetics (Jena & Mackill, 2008), also several markers of DNA matching to main resistance genes have been recognized. The 50 genes for major resistance to blast are identified and a number of resistance genes are created to put heads together broad-spectrum resistance in contrast to pathogen strains tried (Table 1). Nevertheless, maximum rate of this resistance are interrupted in a few years because of novel genotypes of the pathogen can change quickly and overwhelmed host resistance variety (Suh et al., 2009). Incomplete resistance, instead, is typically organized by compound genes, and it can offer an extra steady form of resistance. Relating broad-spectrum resistance genes with compound quantitative resistance genes can be a favorable method to improve durable resistance (Jena & Mackill, 2008; Manosalva et al., 2009). A few *Pi* genes, such as *Pi5*, *Piz-5 (Pi2)*, *Pi40, Pi1* and *Pi9*, have reported to provide a wide-ranging of resistance to *M. grisea* sequesters (Jeon, Chen, Yi, Wang, & Ronald, 2003; Jeung et al., 2007).

Table1. Name of some variety of rice and their resistance genes against blast disease in rice

Variety of Rice (*Oryza sativa L.*)	Different genes against blast disease	Variety of Rice (*Oryza sativa L.*)	Different genes against blast disease
Tjina	*Pib*	**Jinbubyeo**	*Piz*
Tetep	*Pib, Pi1*, and *Pita*	**Junambyeo**	*Pib*
Pai-Kan-Tao	*Pi3*	**IRBL5-M**	*Pi5*

Moroberekan	*Pi5 and Pi7*	**IRBL9-W**	*Pi9*
TKM1	*Piz-t*	**IRBLz5-CA**	*Piz-5*
Tadukan	*Pita*	**IRBLz-Fu**	*Piz*
Zenith	*Piz*	**IRBLzt-T**	*Piz-t*
LAC23	*Pi1*	**IRBLi-F5**	*Pii*
IRBLi-F5	*Pii*	**IRBLb-B**	*Pib*
IRBLz-Fu	*Piz*	**IRBL1-CL**	*Pi1*
IRBLz5-CA	*Piz-5*	**IRBLta-CT2**	*Pita*
IRBL1-CL	*Pi1*	**Ilpumbyeo**	*Pib, Pii*
IRBL5-M	*Pi5*	**Palgongbyeo**	*Pib, Pia*

Conclusion

Some researchers believe that blast disease will be controlled by incorporating several resistance gene into one variety of rice which called gene pyramiding method, also some of them have this opinion that strong resistance variety of rice will be produced by only partial resistance genes (Wang et al., 1994). Managing of rice blast disease is regularly related to chemicals, the normal strategy of mechanism programs needs a suitable accepting of the fungicide resistance wonder in the field of huge pathogens (Wei et al., 2009). Thus it is able to control blast disease by an incorporated organization method with variability of systems –

resistant cultivars, cultural performs and using of chemical– created on the facts from disease estimating structures.

References

Bres-Patry, C., Lorieux, M., Clement, G., Bangraz, M., & Ghesquiere, A. (2001). Heredity and genetic mapping of domestication-related traits in a temperate japonica weedy rice. *Theoretical and Applied Genetics 102*, 118–126.

Cai, W., & Morishima, H. (2002). QTL clusters reflect character associations in wild and cultivated rice. *Theoretical and Applied Genetics, 104*, 1217–1228.

Castilla, N., Savary, S., Cruz, C. M. V., & Hmarch, L. (2010). Rice Blast. *Rice science for a better wrold*, 1-3.

Couch, B. C., & Kohn, L. M. (2002). A multilocus gene genealogy concordant with host preference indicates segregation of a new species, Magnaporthe oryzae, from M. grisea. *Mycologia, 94*, 683-693.

Hajime, K. (2001). Rice blast disease. *The Royal Society of Chemistry, 12*(1), 23-25.

Jena, K. K., & Mackill, D. J. (2008). Molecular markers and their use in marker-assisted selection in rice. *Crop Sci, 48*, 1266-1276.

Jeon, J. S., Chen, D., Yi, G. H., Wang, G. L., & Ronald, P. C. (2003). Genetic and physical mapping of Pi5(t), a locus associated with broadspectrum resistance to rice blast. *Theor. Appl. Genet, 269*, 280-289.

Jeung, J. U., Kim, B. R., Cho, Y. C., Han, S. S., Moon, H. P., Lee, Y. T. (2007). A novel gene, Pi40(t), linked to the DNA markers derived from NBS-LRR motifs confers broad spectrum of blast resistance in rice. *Theor. Appl. Genet, 115*, 1163-1177.

Kim, Y. S., Oh, J. Y., Hwang, B. K., & Kim, K. D. (2008). Variation in sensitivity of Magnaporthe oryzae isolates from Korea to edifenphos and iprobenfos. *Crop Prot, 27*, 1464-1470.

Li, C., Zhou, A., & Sang, T. (2006). Genetic analysis of rice domestication syndrome with the wild annual species, Oryza nivara. *New Phytologist, 170*, 185–194.

Long, D. H., Lee, F. N., & TeBeest, D. O. (2000). Effect of nitrogen fertilizer on disease progress on susceptible and resistant cultivars. *Plant Disease, 84*, 403-409.

Manosalva, P. M., Davidson, R., Bin, M. L., Zhu, X. Y., Hulbert, S. H., Leung, H. (2009). A germin-like protein gene family functions as a complex quantitative trait locus conferring broad-spectrum disease resistance in rice. *Plant Physiol, 149*, 286-296.

Ou, S. H. (Ed.). (1985). *Rice Diseases. Common wealth Mycological institute* (2ed ed. Vol. 6). Kew, Surrey, England: International Rice Reseach Institute.

Suh, J. P., Roh, J. H., Cho, Y. C., Han, S. S., Kim, Y. G., & Jena, K. K. (2009). The Pi40 Gene for Durable Resistance to Rice Blast and Molecular Analysis of Pi40-Advanced Backcross Breeding Lines. *Genetics and Resistance, 99*(2), 243-250.

Thomson, M. J., Tai, T. H., McClung, A. M., Lai, X. H., Hinga, M. E., & Lobos, K. B. (2003). Mapping quantitative trait loci for yield, yield components and morphological traits in an advanced backcross population between Oryza rufipogon and the Oryza sativa cultivar Jefferson. *Theoretical and Applied Genetics, 107*, 479–493.

Uga, Y., Fukuta, Y., Cai, H. W., Iwata, H., Ohsawa, R., & Morishima, H. (2003). Mapping QTLs influencing rice floral morphology using recombinant inbred lines derived from a cross between Oryza sativa L. and Oryza rufipogon Griff. *Theoretical and Applied Genetics, 107*, 218–226.

Vaughan, D. A., Morishima, H., & Kadowaki, K. (2003). Diversity in the Oryza genus. *Current Opinion in Plant Molecular Biology, 6*, 139–146.

Wang, G. L., Mackill, D. J., Bonman, J. M., McCouch, S. R., Champoux, C. M., & Nelson, R. J. (1994). RFLP mapping of genes conferring complete and partial resistance to blast in a durably resistant rice cultivar. *Genetics, 136*, 421-434.

Wei, C. Z., Katoh, H., Nishimura, K., & Ishii, H. (2009). Site-directed mutagenesis of the cytochrome b gene and development of diagnostic methods for identifying QoI resistance of rice blast fungus. *Pest Manag Sci, 65*(12), 1344-1351.

Xiao, J., Li, J., Grandilloa, S., Ahn, S. N., Yuan, L., & DTanksley, S. (1998). Identification of trait-improving quantitative trait loci alleles from a wild rice relative, Oryza rufipogon. *Genetics, 150*, 899–909.

Xiong, L. Z., Liu, K. D., Dai, X. K., Xu, C. G., & Zhang, Q. (1999). Identification of genetic factors controlling domestication-related traits of rice using an F2 population of a cross between Oryza sativa and O. rufipogon. *Theoretical and Applied Genetics, 98*, 243–251.

Chapter 6 : Research Methodology on Adsorption of Heavy Metals by Agricultural Wastes

*Zhang, X.T., Ismail, M. H. S.

HP: 017-3882645, Email: Naturezxt@hotmail.com

Abstract

The bio-adsorbents such as wasted agricultural materials like rice husk, fruit seeds, peel and so on have been widely studied and applied for removal of heavy metals. This review articles deals with not only the basic adsorption background but also the adsorption process involving thermodynamics and kinetics, applications of heavy metals' adsorption by different types of agricultural wastes, Langmuir and Freundlich isotherms compare, etc., because adsorption thermodynamics, kinetics, mechanisms together with the adsorption performance investigations are the future research trends in this field and current research methods in bio-adsorption of heavy metals by agricultural wastes are also described.

Key words: agricultural wastes, bio-adsorption, Langmuir, Freundlich

1. Adsorption background

Adsorption types can be categorized into two groups: physical adsorption and chemical adsorption (Ruthven, 1984). Physical adsorption forms multiple layers of adsorbent molecules including adsorption of dyes or heavy metals by activated carbons (Meshko et al., 2001; Xu and Liu, 2008), and charcoal (Song et al., 2008), etc. and the binding bond is Van de Waals force which is weak, thus the adsorption heat is much less compared with that of chemical adsorption including hazardous organics adsorbed onto metal oxides (Soon et al., 2005) or heavy metals onto cellulose adsorbent usually (zaki et al., 2012), which forms mono-layer of adsorbent molecules and the binding bonds are chemical bonds as ionic bond or chelating bond, etc., so the adsorption heat is always large (usually >2 or 3 times latent heat of evaporation) (Ruthven, 1984).

2. Methodology on current adsorption study

2. 1 Adsorption isotherm

Always one half-empirical equation: Langmuir isotherm and the other empirical equation: Freundlich isotherm are found suitable to describe the relationship between q_e (capital adsorbate adsorbed at equilibrium, mg/g) and C_e (concentration of adsorbates remained at equilibrium, mg/L). These two models are available for adsorption of gases onto solid surface or solutes in diluted solution onto solid surface (**Motoyuki, 1990**).

2.1.1 Langmuir isotherm

For Langmuir equation, it is written as: $\dfrac{1111}{qbQCQ}$ —— —— (Borah et al., 2008), and whereas Q_m is the maximum capital adsorption capacity of an adsorbent, and it is intensive quantity relating only to the temperature, mutual reaction between adsorbate and adsorbent, characteristics of adsorbent itself, or other variables which are not corresponding to quantitative change like mass change of adsorbent for example. The constant b is called

Langmuir adsorption equilibrium constant and equals k_a/k_d (Ruthven, 1984). K_a is the adsorption rate constant and K_d is the desorption rate constant.

In Langmuir adsorption theory, molecules are adsorbed at a fixed number of well-defined localized sites (active sites). Each site can hold one adsorbate molecule. All sites are energetically equivalent (Langmuir, 1916). Finally, there is no interaction between molecules adsorbed on neighboring sites. So Langmuir isotherm is suitable for representing chemisorptions on a set of distinct localized adsorption sites (Ruthven, 1984).

2.1.2 Freundlich isotherm

For Freundlich equation, it is written as: $\log q_e = \log K_f + \frac{1}{n} \log c_e$ (Borah et al., 2008) and that K_f and n are Freundlich constants and do not have any physical meaning because they are totally empirical. Langmuir and Freundlich isotherm models are frequently used isotherm models for describing the short term and mono component adsorption of metal ions by different materials (Aksu et al., 1999; Yu et al., 2001).

Usually, Freundlich model always better describes physic-sorption (e.g. by activated carbon) of low solute concentration (Yu and Li et al., 2007; Sud, 2008). At low solute concentration with low θ, multiple layer physical adsorption becomes single layer, which is the reason why in short-term, mono component sorption system, Freundlich isotherm also fits the chemical adsorption process sometimes, but not as well as Langmuir model does.

2.2. Adsorption equilibrium constant

If adsorption process fits Langmuir model, since b is Langmuir equilibrium constant, theoretically, it is used as true adsorption equilibrium constant if chemical adsorption process is dominant in a diluted solution system, but b is not true adsorption equilibrium constant under all adsorption conditions (linearity of Langmuir model degrades in concentrated solution system) (Ye, 2010), for Langmuir isotherm is constructed on certain assumptions which are mentioned above, thus the use of b as equilibrium constant is depending on the situation. In some journals, distribution coefficient K_d was found used as a more general

adsorption equilibrium constant (Arivoli, 2009; Connel, Dywer et al., 2005). For K_d is

defined as: $K =\dfrac{a_s}{a_e} = \dfrac{\gamma_s}{\gamma_e}\dfrac{q_e}{C_e}$ and a_s is the activity of solute adsorbed (mg/g); a_e is the

activity of solute in bulky solution (mg/L); γ_s, γ_e are all activity coefficients. When in diluted

sorption system, $C_e \to 0$, $q_e \to 0$, and $\gamma_s \to 1$, $\gamma_e \to 1$, thus K_d equals to $\dfrac{q_e}{C_e}$. Usually, K_d is

obtained by plotting $\ln(q_e/C_e)$ versus C_e, and extrapolating C_e to zero (Khan and Singh,

1987), then the exponential of intercept is K_d. It can be found that K_d does not have apparent

relationship with active sites percentage θ, but, if compare K_L with K_d by changing the form

of Langmuir equation, it has:

$$\dfrac{q_e}{C_e} = \dfrac{Q_m K_L}{1 + K_L C_e}$$

$$\therefore \left.\dfrac{q_e}{C_e}\right|_{C_e \to 0} = Q_m K_L \tag{1}$$

Thus $K_d = Q_m \cdot K_L$ (Ye, 2010), and Q_m is related to θ. As a more general adsorption

equilibrium constant, K_d is deduced based on the equilibrium of sorbate's chemical potential

separately in liquid and solid phase, can be adopted to calculate adsorption enthalpy, entropy

conveniently regardless of adsorption type.

2.3. Adsorption Mechanism

The removal of metal ions from aqueous streams using agricultural materials is based on

metal bio-sorption (Volesky and Holan, 1995). The bio-sorption involves a solid phase

(adsorbent) and liquid phase (solvent) containing dissolved particles (adsorbate), so it is a

cross phase reaction. Due to the characteristics of agricultural wastes adsorbent and the

affinity between surface active sites and adsorbates, the adsorption process is complex

affected by several mechanisms involving chemisorptions (ion exchange,

complexation/chelating, etc), physic-sorption (by physical force like electrostatic force),

surface and pore adsorption, entrapment in inter and intra-fibrillar capillaries and spaces of

the structural polysaccharides network as a result of the concentration gradient and diffusion through cell wall and membrane (Basso et al., 2002; Sarkanen and Ludwig, 1971; Qaiser et al., 2007).

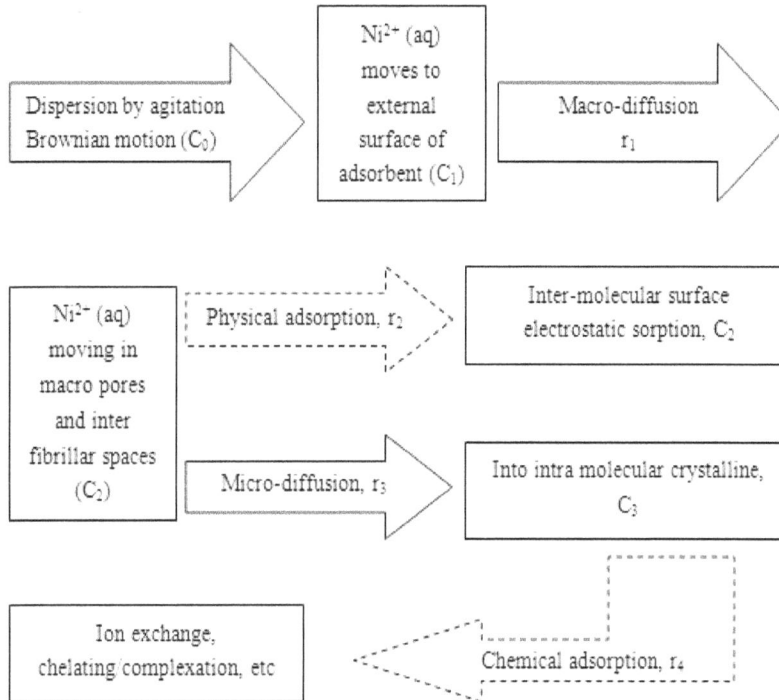

Note: r_i — rate at different steps, C_i — concentration of Ni^{2+} at different steps

Figure 1 Mass transfer flow of Ni^{2+} in bio-adsorption system

2.4. Adsorption kinetics

Numerous kinetic models have described the reaction order of adsorption systems. These include first-order and second-order reversible ones, and first-order and second-order irreversible ones, pseudo-first-order and pseudo-second-order ones (Sud, 2008).

Some studies showed that Lagergren's pseudo-second order equation well correlated the Cu^{2+} adsorption on chelating cellulose adsorbent (Connel, Dywer et al., 2005) and on chemically treated orange peel (Guo and Liang, et al., 2008), while Lagergern's pseudo-first order kinetics was suitable for Ni (II), Co (II), Cd (II) and Zn (II) adsorption description on orange peel cellulose adsorbent (Li and Tang, 2008).

Lagergren's pseudo-first order equation can be expressed as: $\dfrac{dq_t}{dt}$ 1 and

Lagergren's pseudo-second order equation is —— $= kq_t q_{et}$ 2.

2.5. Adsorption thermodynamics

For adsorption process, if pressure is constant and so is the volume and temperature, then,
adsorption reaction heat $Q_r = \Delta H_r$. $_{rm}$ ^{o} can be achieved by Van't Hoff isochore:

$$\frac{\Delta_{rm} H}{RT}$$

. It is only valid if $_{rm}$ ^{o} varies little (G.K. Vemulapalli, 1993) with

temperature change. K_e is adsorption equilibrium constant.

Standard Gibbs energy can be calculated through equation: $\Delta G = -RTK$ and standard

entropy is calculated by $\Delta S = RTK$ ——————— ln ———— .

2.6. Application of bio-adsorbents

Adsorption by activated carbon is of high cost and its adsorption capacity loss during
regeneration restricts its application. Recently attention has been diverted towards the
biomaterials which are byproducts or the wastes from large scale industrial operations and
agricultural waste materials (Sud, 2008). The major advantages of bio-sorption over
conventional removal of metal ions treatment methods include: low cost, high efficiency,
minimization of chemical or biological sludge, no additional nutrient requirement, and fine
regeneration of bio-sorbents and possibility of metal recovery.

Studies revealed various materials like crop husks, hulls, fruit fresh or peels, coconut shells,
etc., as adsorbents with or without chemical treatments, but most of those adsorbents are
made from one part of botanical bio-mass

Different studies also revealed the efficiency of removal of different aqueous heavy metal
ions by different mono-substance botanical adsorbents. For Cr (VI), efficiency reached 100%

by beech saw dust (Acar and Malkoc, 2004); 100% for Pb (II) by Coconut char based activated carbon (Gajghate et al., 1991); 98% for Cd (II) by powder of green coconut shell (Pino et al., 2006); 99~100% by Cassia fistula biomass for Nickel (Hanif et al., 2007), etc.

2.7. Chemical treatment of bio-adsorbent

The way by introducing adsorption functional group to enhance the adsorption ability of adsorbent is widely used for many bio-adsorbents. Those functional groups including hydroxyl, carboxyl, carbonyl, sulfo, amino, cyanoethyl, etc, adsorbents modified by functional groups introduced thus can be categorized into cation, anion, or amphoteric ones (Jang and Huang, 2008). For cation adsorption adsorbents, always carboxyl, sulfo or phosphoryl is introduced by etserification through organic or inorganic acid reaction with hydroxyls in cellulose of botanical bio-adsorbents (Jang and Huang, 2008). Also, it was reported that alkalization of cellulose to prepare cellulose salt was conducted first for promoting the next esterification process (Li & Tang et al., 2008).

2.8. Desorption method

Through ion exchange by adding acid onto adsorbent can efficiently recover the spent adsorbent. Studies showed that spent chemically treated orange peel adsorbent for copper adsorption can be used for 5 times with efficiency changing from 97.38% to 92.03% after being soaked in 0.1 M HCl for 3 hours (Guo and Liang, et al., 2008).

Desorption experiments (Li and Tang et al., 2008) showed that for Nickel ion desorption of phosphoric acid modified orange peel in 0.1 M HCl, the recovery efficiency of metal ions reached 80.11% which was the highest among the efficiencies resulted by other HCl concentrations ranging from 0.00~0.20 M with interval of 0.05M.

3. Conclusion

Nowadays, the research on heavy metal adsorption by agricultural wastes shifts from material selection, operation parameters optimization to more fundamental adsorption mechanisms study as transportation phenomena, adsorption kinetics and thermodynamics investigation, and adsorption constant determination under different solution concentration. Modifications of existed or new methods to explore the adsorption of heavy metals on/into agricultural wastes adsorbents are continuously given, which will definitely be helpful for us to understand and apply this technology.

Reference

Zaki, A. A., EL-Zakla, T., Geleel, M. A. E. Modeling kinetics and thermodynamics of Cs$^+$ and Eu^{3+} removal from waste solutions using modified cellulose acetate membranes. Membrane Science, In Press, Accepted Manuscript, Available online 3 January 2012.

Acar, F. N., Malkoc, E. (2004). Removal of Chromium (VI) from aqueous solution by Fagus orientalis. Bioresources.Technology, 94, 13–15.

Basso, M. C., Cerrella, E. G., Cukierman, A. L. (2002). Lignocellulosic materials as potential biosorbents of trace toxic metals from wastewater. Chemical Research 41, 3580–3585.

Borah, D., Satokawa, S., Kato, S., Kojima, T. (2008). Surface-modified carbon black for As (V) removal. Journal of Colloid and Interface Science, 319, 53–62

Connell, D. W., Birkinshaw C., Dwyer T. F. (2005). A chelating cellulose adsorbent for the removal of Cu (II) from aqueous solutions. Journal of applied polymer science, 99(6), 2888~2897.

Feng, N. C., Guo, X. Y., Liang, S., Tian, Q. H. (2008). Biosorption of Cu (II) ion on modified orange peel. *The Chinese journal of no ferrous metals*, 18(z1).

Gajghate, D. G., Saxena, E. R., Vittal, M. (1991). Removal of lead from aqueous solution by activated carbon. *India Journal of Environment Health* 33, 374-379.

Hanif, M. A., Nadeem, R., Zafar, M. N., Akhtar, K., Bhatti, H. N. (2007). Nickel (II) biosorption by *Casia fistula* biomass. *Journal of Hazard.Material B* 139, 345–355.

Jiang, Y., Huang C. J., Pang H., Liao B. (2008). Progress in Cellulose-based Adsorbents. *Chemistry*, 71(12).

Juang, R. S., Jiang, J. D. (1995). Recovery of nickel from a simulated electroplating rinse solution by solvent extraction and liquid surfactant membrane. *Journal of Membrane Science*, 100, 163-170.

Khan, A. A., Singh, R.P. (1987). Adsorption Thermodynamics of Carbofuran on Sn (IV) Arsenosilicate in H^+, Na^+ and Ca^{2+} Forms. *Colloids and surface*, 24, 33-42

Langmuir, I. (1916). The constitution and fundamental properties of solids and liquids. *Journal of the American Chemical Society* 38(11): 2221–2295.

Li, X. M., Tang, Y., Cao X. J. et al. (2008). Preparation and evaluation of orange peel cellulose adsorbents for effective removal of cadmium,zinc, cobalt and nickel. *Colloids and surface A: Physicochem. Eng. Aspects*, 317, 512-521.

Meshko, V., Markovska, L., Mincheva, M., Rodrigues, A. E. (2001). Adsorption of basic dyes on granular acivated carbon and natural zeolite. *Water Research*, 35, (14), 3357-3366.

Motoyuki, S. (1990). *Adsorption Engineering*. Kodansha Ltd, Tokyo and Elesvier science publishers B.V., Amsterdam.

Periasamy, K., Namasivayam C. (1995). Removal of Nickel (II) from aqueous solution and nickel plating industry wastewater using an agricultural watse: peanut hulls. *Waste management*, 15(1), 63-68.

Pino, G., de Mesquita, L., Torem, M., Pinto, G. (2006). Biosorption of heavy metals by powder of green coconut shell. *Separation Science & Technology*. 41, 3141–3153.

Qaiser, S., Saleemi, A. R., Ahmad, M. M. (2007). Heavy metal up take by agrobased waste materials. *Environment.Biotechnology*, 10, 409–416.

Ruthhven, D. M. (1984). *Principles of adsorption and adsorption processes*. John Wiley & Sons, Inc., Canada.

Sarkanen, K. V., Ludwig, C. H. (1971). *Lignins-Occurance, Formation, Structure and Reactions*. Wiley-Interscience, NewYork, 1.

Song, Y. W., Ming, H. T., Sheng, F. L., Ming, J. T. (2008). Effects of manufacturing conditions on the adsorption capacity of heavy metal ions by Makino bamboo charcoal. *Bioresource Technology*, 99(15), 7027-7033.

Soon, K. J., Takahashi, A., Masayuki, M., Cai, R., Kiminori, I. (2005). Chemical adsorption of phosgene on TiO_2 and its effect on the photocatalytic oxidation of trichloroethylene. *Surface Science*, 598(1–3), 174-184.

Sud, D., Mahajan G., Kaur M.P. (2008). Agricultural waste material as potential adsorbent for sequestering heavy metal ions from aqueous solutions – A review. *Bioresource technology*, 99, 6017-6027.

Xu, T., and Liu, X. Q. (2008). Peanut Shell Activated Carbon: Characterization, Surface Modification and Adsorption of Pb^{2+} from Aqueous Solution. *Chinese Journal of Chemical Engineering*, 16(3), 401-406.

Vemulapalli, G.K. (1993). *Physical chemistry*. **Prentice Hall, Englewood Cliffs, New Jersey.**

Vijayakumaran, V., Arivoli, S., Ramuthai S. (2009). Adsorption of Nickel Ion by Low Cost Carbon-Kinetic, Thermodynamic and Equilibrium Studies. *E-journal of chemistry*, 6, 347-357.

Volesky, B., Holan, Z. R. (1995). Biosorption of heavy metals. *Biotechnology Progress*, 11, 235–250.

Ye, S. L. (2010). The limiting definition of adsorption equilibrium between heterogeneous solid-liquid phase and its relationship with covering degree θ. *Environmental Chemistry*, 29(4).

Yu, C., Li, X. H., Qiu, J. S., Sun Y. F. (2007). Removal of sulfur-containing compounds from oil by activated carbon adsorption. *Journal of fuel chemistry and technology*, 35(1).

Zhao, M., Duncan, J. R. (1998). Removal and recovery of nickel from aqueous solution and electroplating rinse effluent using Azolla filiculoides. *Process Biochemistry*, 33(3), 249-255.

Chapter 7: Research Methodology on Continuous & Non-continuous Chromatographic Separation of Protein

*Zhang, X.T., Ismail, M. H. S.

HP: 017-3882645, Email: Naturezxt@hotmail.com

Abstract

Chromatography separation has been widely applied for isolation and purification of many compounds, including many synthetic or natural bio-molecules. Proteins and polypeptides are now easily identified and analyzed with the help of chromatography development. This article gives a review on the non-continuous lab-scale purification and the large scale continuous separation techniques and equipments for proteins or polypeptides by ion exchange, reversed phase, gel permeation liquid chromatography. Operation parameters are not discussed in detail here, but more on the process design and columns selection.

Key words: large scale chromatography, protein & polypeptide separation, ion exchange, reversed phase, gel permeation

1. General chromatographic adsorption and separation

Chromatography separation method has long been used for separate and analyzes many compounds especially for organics and it becomes more efficient and convenient after modified to be operated under high pressure with liquid as mobile phase like HPLC and FPLC (Lindsay and Kealey, 1987), reducing the residual time and dragging force in highly packed beads column which can gives a higher resolution.

Basically, Chromatography technologies can be divided into ion exchange, size exclusion (gel filtration chromatography), revered phase (RP) and affinity chromatography. For protein and polypeptides separation, usually ion exchange columns are used because proteins and polypeptides have different surface net charge under different pH of eluents and buffers

(Wikipedia), and by change the composition ration of eluent, a changing pH or polarity gradient can be achieved to raise the resolution of separation of proteins and polypeptides. Compared to affinity separation, ion exchange is relatively economic though affinity chromatography gives a more special binging to highly increase the purity of target products (Waugh, 2005). Ion exchange can also help when the molecular weights of different proteins or polypeptides are similar to each other where a gel chromatography fails (Dauphin et al., 2007). But usually to separate the bio-targets from the raw material such as broth is not so easy due to great amount of other interfering unwanted proteins, polysaccharides, lipids, etc. So many different types of ion exchange and other types of columns in series and pre-treatments before chromatographic separation are commonly adopted , (Kim, Kim & Lim et al., 2009; Shi & Li et al., 2006; Steiner, Knecht & Gruetter, 1990) but it is uneconomic, time costing and inconvenient to be enlarged for industrial scale, also for those proteins which render similar surface charge under same pH, the efficiency of ion exchange chromatography is thus reduced, so usually pre-treatments of raw materials are needed and pre-qualitative analysis of the proteins inside are done to further choose the correct ion exchange or other chromatography columns to greatly reduce the steps involved, meanwhile new designs on column mechanical and separation process including conditioning, washing, elution, regeneration are also studied to modify the protein/ polypeptides ion exchange chromatography separation and purification in continuously industrial large scale.

1.1 Non-continuous chromatographic separation

For non-continuous ion exchange chromatography, always a fixed bed and column or fixed beds and columns in series can be found. To better utilize this technology, steps reduction or columns numbers reduction is strongly needed, but as a whole, fixed bed technology is no suitable for large scale production but small scale ones like laboratory separation and purification with small quantity yield. HPV16L1 protein expressed by *Saccharomyces cerevisiae* Y2805 was reported successfully recovered around 60%, the highest so far, by two steps chromatographic separation after the broth was firstly treated by ammonium sulfate to remove contaminants (Kim, Kim & Lim et al., 2009). The pre-treated sample was first eluted in a heparin column which is regarded a cation exchanger with affinity binding ability, and then the eluant went through the second column packed of P-11 phosphocellulose cation-exchange resin. The key to obtain only two steps ion exchange separation lies in the effective pre-precipitation of many contaminants including those unwanted proteins which have similar surface net charge or binding affinity.

Also, Shi and Li et al. separated and purified recombinant Hirudin Variant 2 (HV2) and its two truncated derivatives expressed in *Pichiapastoris* by two steps ion exchange chromatography (Shi & Li et al., 2006). Before applied the broth to the columns, it was ultra-filtrated by the membranes with 80 and 3 KD cut-off and the retentant was then first applied to cation-exchange chromatography on SP-Sepharose Fast Flow resin, after which active fraction was collected and diluted with five-fold volume. Half of the diluted active fraction consequentially went through the second anion-exchange column 15Q with gradient elution by NaCl solution. The final active fraction was desalinated by SephadexG-25 and analyzed by Syn Chropak C4 RP-HPLC. Shi et al. provided an alternative for usual reversed phase chromatography by a two-step ion exchange, but did not compare the efficiency of them and obviously poses a more step of desalting though separating the final target from the eluant is always unavoidable for ion exchange chromatography.

Noticeable, Steiner et al. developed a five steps isolation and purification process involving gel permeation, ion exchange and RP chromatography for Hirullin P6 and P18 (Steiner, Knecht & Gruetter, 1990). The crude extract first passed through Sephadex G50 gel permeation column and then applied to a Q Sepharpse fast flow anion column to remove many unwanted peptides. The active fraction eluted by NaCl was not desalted before being applied into next C18 RP-HPLC which plays a role for desalting and two peaks representing P6 and P18 were showed though with low resolution. So, Steiner et al. further applied the pooled P6 and P18 eluted from C18 to Phenyl silica RP-HPLC which gives a higher resolution about the two peaks while introduced the interfere of TFA salts since TFA was one component in the eluent. At last, a Mono Q anion column was used to remove those salt contaminants as a final purification. The total yield at last is 0.008% for P6 and 0.003% for P18. It can be seen that the isolation and purification process involved too many steps which gave a very low yield. Actually the isolation steps aimed to remove majorly contaminants can be replaced by membrane filtration or ultra-filtration and then pass the active parts into Phenyl silica RP-HPLC first followed by Sephadex G-25 for desalting or by ultra-filtration again, thus the number of different columns needed will be largely reduced.

1.2 Continuous chromatography and chromatographic adsorption

In order to enlarge the scale of separation and purification of protein or peptides, continuous separation columns are very useful. One kind of continuous chromatography separation technology is based on the movement of columns, or simulated by changing the feeding and elution valves for those fixed chromatographic beds whose solid particles (stationary phase) are relatively static to the container. So it is called continuous fixed bed chromatography. Usually, counter-flow or cross-flow is applied in such continuous system to maximize the separation efficiency of or simultaneously separate the two components and co-current is rarely seen and gives a same efficiency with batch process unless a recycle is involved (Gordon, Moore & Cooney, 1990). Highly packed solids in the column raise the resolution while cause the problem of pressure drop and clotting. One simplest approach to realize a counter-flow continuous separation is called Rotating Column Chromatography by rotating serial inter-connected columns in a circular array against the direction of feeding flow, but this design has leaking problems and not available for commercial use (Barker, 1971). Andersson et al. designed another kind of rotating column cation-exchange chromatography composed of 20 columns for Lactoperoxidase and Lactoferrin separation from whey protein. That 20 columns change their positions in turn to play roles for conditioning, washing, elution, regeneration in a disconnected series to each other with a valve down head on each column corresponds to each fixed valve upper head mounted on central multi-port system which coordinates the liquid flow into and out of the columns (Andeesson & Mattiasson, 2006).

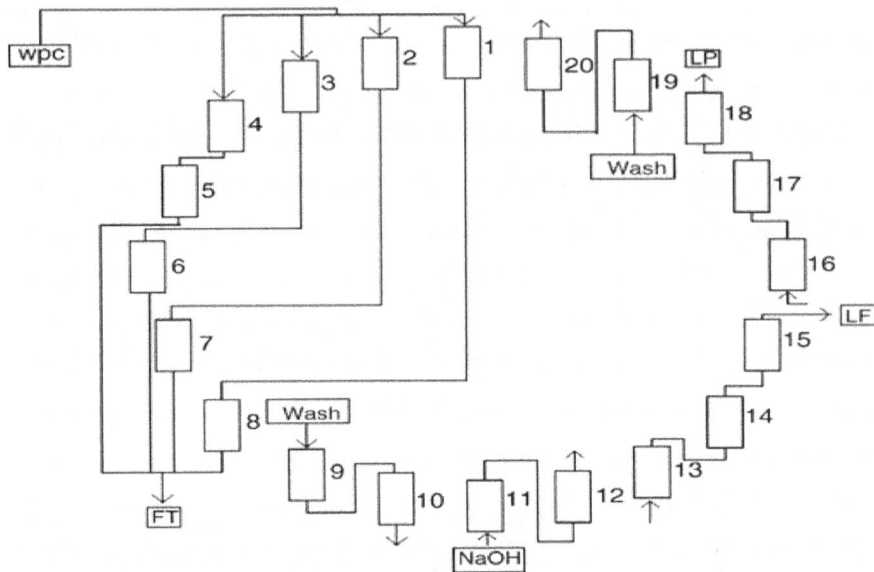

Fig1. Schematic diagram of 20 columns system (Source: Andeesson & Mattiasson, 2006)

The elution is counter-flow against the loading and the results showed that 48% rise in productivity, 6.5 times decrease in buffer consumption and 4.8 times higher target protein concentration compared to traditional fixed 20 columns in parallel.

Rather than move the columns, simulated strategy is to alternatively move the loading and elution valves instead. Such Simulated Moving Bed system (SMB) is more industrially scalable and can collect the targets simultaneously, but shares the problem with packed bed operation (Gordon, Moore & Cooney, 1990). By switch the valves for loading or elution, two mixed the components will immigrate to the opposite direction due to their difference in immigration speed compared to the simulated movement of the bed relates to the period of valve switch. But for a three components system, simple valve change cannot separate all the three components at the same time because there always is one component which has an immigration speed between the other two's and that causes the overlap for fraction collection. So a cut-off before the valve switch is applied to stop the circulation into the column and it operates like a non-moving fixed bed column to first separate out one component and then switch off the feeding valve to only let the elution valve move to further separate the rest two like a simulated moving bed column. It is called new JO system (*ORGANO Corporation*).

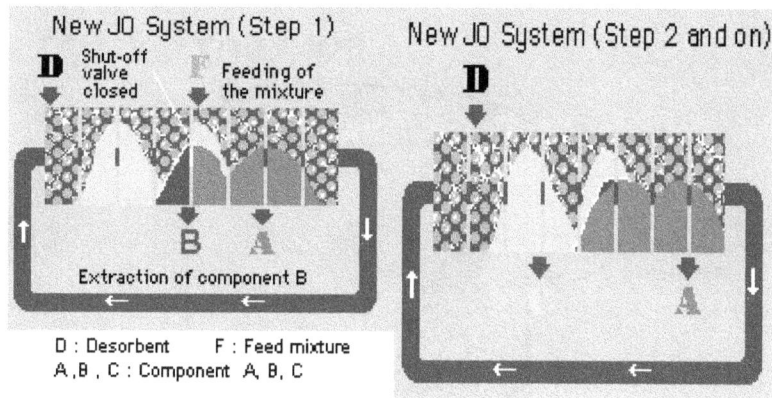

Fig2. New JO system (Source: *ORGANO Corporation*)

For more than three target components need to be separated simultaneously, always simulated moving beds in series are needed, which increases the separation steps and lowers the yield.

Another way to simultaneously collect different targets at each effluent port is to run the continuous separation in a Rotating Annular Chromatography (RAC). This idea came from 60 years ago (Martin, 1949). Due to both gravitational and centrifugal force, and the difference in affinity to the packed solids, the traces of components are helical and end up at different effluent positions. During the continuous process, the feeding flows in at one point and the eluent is pouring down through the whole annulus, thus gradient elution is hard to achieve because of back-mixing, so this technology is better for gel filtration chromatography. Fox et al. purified the Myoglobin and components of skim milk by RAC on Saphadex G-25 gel permeation chromatography (Fox, 1969; Fox & Calhoun et al., 1969; Nicolas & Fox, 1969). The capacity and running mode of GAC is similar to conventional batch columns if it is not applied for gel chromatography, also leaking and design difficulties exist.

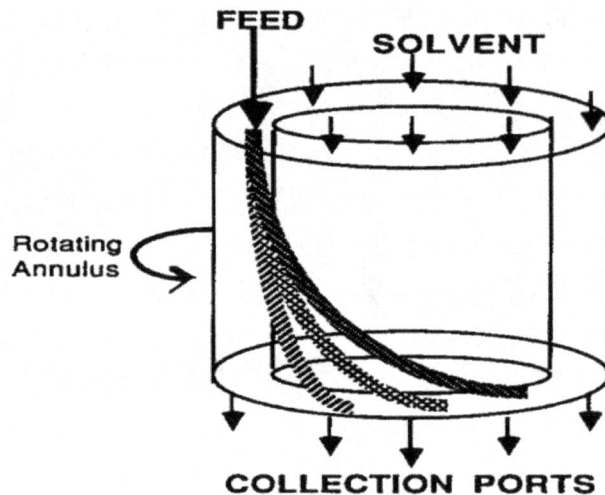

Fig3. Rotating Annular Chromatography (Source: Gordon, Moore & Cooney, 1990)

Other than moving columns or simulated moving bed chromatography, real bed moving chromatography technologies are also very common now. Those include packed bed, fluidized bed, CSTR and adsorption/ filtration types (Gordon, Moore & Cooney, 1990).

Continuous Packed Bed or Plug Flow Moving Bed (PFMB) was early introduced by Oak Ridge National Laboratory. A number of inlets and outlets (for conditioning, loading, washing, elution, and regeneration) are along the running column and in which is the packed resin prevented to be fluidized by hydraulic ram. The packed resin will be cycled through the outside loop by a jet at the bottom of the column. Because the packed resin moves slowly thus causes little pressure drop to the counter-flow of loadings, but due to its highly complex in mechanical design and strict requirement in operation balance, it inclines to fail.

Selke and Bliss constructed a fluidized counter-flow moving bed for ion exchange chromatography (Selke & Bliss, 1951).

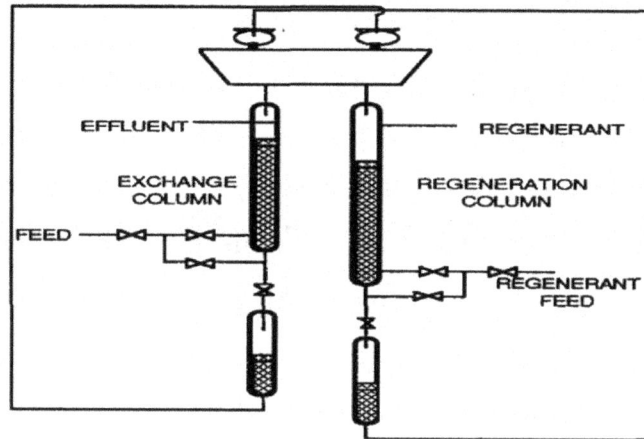

Fig4. Continuous fluidized ion exchange columns (Source: Gordon, Moore & Cooney, 1990)

Such system is composed of exchange and regeneration regions, in which axial mixing cannot be avoided but solids clotting can be prevented as the applied broth always contains many cell's debris, so an integrated system of filtration before the entrance of fluidized bed column to block large sizes of particles and at the rear of column to hold the resins is developed by Bartels et al. to extract the antibiotic, streptomycin (Bartels & Kleinman et al., 1958). In order to resolve the axial mixing problem, Olin et al. added a concentric cylinder in which the inner cylinder rotated at a high speed to maximize the radical mixing but channeling mixing like a plug reactor. And Cloete and Streat introduced concept to divide the fluidized bed into different regions by perforated plates in analog to gas adsorption tower (Cloete & Streat, 1963). Different regions can carry out tasks for conditioning, loading, washing, elution, and regeneration alternatively, thus realizing continuous separation.

A more delicate approach is to prevent the back mixing though magnetic field which aligns the magnetic adsorbents moving forward against the loading flow and absorb the proteins/ polypeptides by ligands in first satge and then desorb the adsorbents by desorbing solvent in the second stage (Bums & Graves, 1985; Siegell & Dupre et al., 1986). Such technology is promising for bio-products separation but expensive which lays difficulties for industrial application.

Continuous Stirred Tank Reactor (CSTR) is usually used for waste water treatment and chemicals synthesis, but can be used to separate proteins or polypeptides if combined with filtration technology. Mattiasson developed a continuous counter-flow affinity adsorption in CSTR with ultra-filtration by using soluble polymers as adsorbent (Matfiasson & Ling et al., 1983; Matfiasson & Ramstorp, 1983; Matfiasson & Ramstorp, 1984). When the certain target is absorbed onto the polymer, the size of polymer changes and thus can be retained by the membrane has a similar cutoff size to the polymer, while the other unwanted components of smaller size will pass through. The adsorption, washing, elution and regeneration steps are conducted in separate vessels continuously. But this technology is vulnerable to solid contaminants whose sizes are large or similar to the polymer-protein particle, so more CSTRs are designed to remove the incoming contaminants. In the first stirred adsorption stage, a washing stream is used to bring away most contaminants in the crude feed while the targets are adsorbed on to the adsorbent which are retained by the micro-porous metal screen (Gordon & Cooney, 1990; Gordon & Tsujimura et al., 1990). The elution stages are multiple and eluants are selectively collected and concentrated. Regenerated adsorbents from the elution steps are pumped back to adsorption stage.

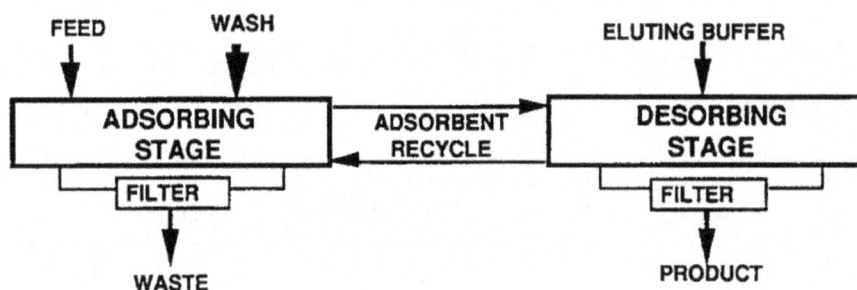

Fig5. Continuous Affinity Recycle Extraction (CARE) (Source: Gordon, Moore & Cooney, 1990)

The integration of CARE to downstream process was demonstrated for both ion-exchange and affinity adsorption for purification of enzyme 13-galactosidase from partially clarified E.colilysate feed with high product yield.

Except to flow the adsorbents, move the equipment contains the adsorbents is another approach to realize a continuous separation process. Porath and Sunndberg Suggested to enclose the affinity adsorbents in a "tea-bag" like membrane bag whose pore sizes are only permeable to the target proteins (Porath & Sundberg, 1972). The process is continuous since the bags are transported on a moving belt and suspended in different vessels for adsorption, washing, elution and regeneration. But the adsorption kinetics is very slow due to long time diffusion of the targets into the bag and bypass of fluid may happen even agitation is provided (Gordon, Moore & Cooney, 1990). The validity of the bag adsorption in crude feeds was not demonstrated since the relevant experiments were tested in "clean" system (Gordon, Moore & Cooney, 1990).

2. Conclusion

Though lab-scale and preparative scale of protein or polypeptides are already mature, but the effort to scale up the separation and purification of them is still need though there already existed certain large scale continuous separation chromatographic equipments, but the deficiencies are obvious and some are already abandoned due to complex design or operation requirements. Some conceptual scheme as magnetically aligned fluidized bed as example is promising, but lacks of experimental proofs. When it concerns industrial production, economic factors cannot be ignored, either. Affinity column are usually expensive, but very special to the target. So the development of flexible continuous chromatographic separation and purification still is on its way.

Reference

Andersson, J., Mattiasson, B. (2006). Simulated moving bed technology with a simplified approach for protein purification: Separation of *lactoperoxidase* and *lactoferrin* from whey protein concentrate. *Journal of Chromatpgraphy A*, 1107, 88-95.

Barker, P. E. (1971). *Continuous chromatographic refining, in Progress in Separation and Purification* (Ed). Perry, et al., Wiley-Interscience.

Bartels, C. R., Kleinman, G., Korzum, J. N. and Irish, D. B. (1958). A novel ion-exchange method for the isolation of Streptomycin. *Chemical Engineering Progress*, 54, 49-51.

Bums, M.A. and Graves, D. J. (1985). Continuous affinity chromatography using a magnetically stabilized fluidized bed. *Biotechnology Progress*, 1, 95-103.

Cloete, F.L.D. and Streat, M. (1963). A new continuous solid fluid contacting technique. *Nature*, 1199-1200.

Dauphin, P., Morgan, Stephanie., Agorio, M., Kandiyoti, H. (2007). Probing Size Exclusion Mechanisms of Complex Hydrocarbon Mixtures: The Effect of Altering Eluent Compositions. *Energy & Fuels*, (6)21, 3484–3489.

Features of the New JO Chromatographic Separation System. *ORGANO Corporation.* Retrieved from http://www.organo.co.jp/technology/hisepa/en_hisepa/newjo/jo1.html

Fox, J. B. Jr. (1969). Continuous chromatography apparatus: II, Operation. *Journal of Chromatography*, 43, 55-60.

Fox, J. B. Jr., Calhoun, R. C. and Eglinton, W. J. (1969). Continuous chromatography apparatus: I., Construction. *Journal of Chromatography*, 43, 48-54.

Gordon, N. F. and Cooney, C. L. (1990). Impact of continuous affinity-recycle extraction (CARE) in downstream processing in Protein Purification from Molecular Mechanisms to Large-Scale Processes. *ACS Symposium Series*, 427, 118-138.

Gordon, N., Moore, C. M. V. and Cooney, C. (1990). An overview of continuous protein purification processes. *Biotechnology Advances*, 8, 741-762.

Gordon, N.F., Tsujimura, H. and Cooney, C. L. (1990). Optimization and simulation of continuous affinity-recycle extraction (CARE). *Bioseparation*, 1, 9-21.

Ion chromatography. *Wikipedia.* Retrieved from http://en.wikipedia.org/wiki/High-performance_liquid_chromatography#Bioaffinity_chromatography

Kealey, D., and Lindsay, S. (1987). *High performance liquid chromatography.* John Wiley and Sons, New York, NY .

Kim, H. J., Kim, S. Y., and Lim, S. J.et al. (2009). One-step chromatographic purification of human papillomavirus type 16L1 protein from *Saccharomyces cerevisiae. Protein Expression and Purification*, 70, 68-74.

Martin, A.J.R. (1949). Summarizing paper, Discuss. *Faraday Society*, 332-336.

Matfiasson, B., Ling, T. G. I. and Nilsson, J. L. (1983). *Ulttaffitration affinity purification, in Affinity Chromatography and Biological Recognition*, (Ed). Chaiken, et al. Academic Press, 223-228.

Mattiasson, B. and Ramstorp, M. (1983). Ultrafiltration affinity purification. *Annual of NY Academic Science*, 413, 307-309.

Mattiasson, B. and Ramstorp, M. (1984). Ultraffltration affinity purification; isolation of concancavalin A from seeds of Canavalia Ensiformis. *Journal of Chromatography*, 283, 323-330.

Nicholas, R.A. and Fox, J. B. Jr. (1969). Continuous chromatography apparatus – III, Application. *Journal of Chromatography*, 43, 61-65.

Porath, J. and Sundberg, L. (1972). Specific extraction of enzymes using sold surfaces. In (Ed), *The Chemistry of Biosurfaces, Hair* (pp.633-661). Marcel Dekker, Inc.

Selke, W.A. and Bliss, H. (1951). Continuous Countercurrent Ion Exchange. *Chemical Engineering Progress*, 47, 529-533.

Shi, B. X., Li, J. C., et al. (2006). Two-step ion-exchange chromatographic purification of recombinant hirudin-II and its C-terminal-truncated derivatives expressed in *Pichia pastoris*. *Process Biology*, 41, 2446-2451.

Siegell, J. H., Dupre, G. D. and Prikle, J. C. Jr. (1986). *Chromatographic separations in a crossflow magnetically stabilized bed, in Recent Advances in Separation Techniques - III*, (Ed). Li. AIChE Symposium Series, 250(82), 128-134.

Steiner, V., Knecht, R., and Gruetter, M., (1990). Isolation and purification of novel hirudins from the leech *Hirudinaria manillensis* by high-performance liquid chromatography. *Journal of Cromatography, Bomedical Aplications*, 530, 273-282.

Waugh, D. S. (2005). Making the most of affinity tags. *TRENDS in Biotechnology* 23(6), 316-320.

Chapter 8: A REVIEW OF PM$_{10}$ TREND FOR THE INDOOR AIR QUALITY PARAMETERS AND ITS EFFECT TOWARDS CHILDREN

[a] Tezara, C., N.M. Adam [a]*, Juliana, J. [b]*, and Mariani, M. [c]*.

[a] Department of Mechanical and Manufacturing Engineering, Universiti Putra Malaysia, 43400 Serdang, Selangor, Malaysia

[b] Department of Community Health, Universiti Putra Malaysia, 43400 Serdang, Selangor, Malaysia

[c] Department of Ecology, Universiti Putra Malaysia, 43400 Serdang, Selangor, Malaysia

Abstract

Particulate matter - 10 (PM$_{10}$) is a type the air pollutants that is present wherever people live. This paper reviewed about the concentration levels of PM$_{10}$ in area where the children spend their time and its effect on their health. This paper reviewed the study on PM$_{10}$ using different equipments such as DustTrak and Caselle Microdust Pro and etc. The result reported the indoor PM$_{10}$ and outdoor PM$_{10}$ concentrations in children environment. The indoor PM$_{10}$ level is 2-3 times higher than outside level due to the occupant density and movement, activity such as cooking and building characteristic. The PM$_{10}$ concentration levels in different areas like city and suburb did not show any significant difference though in the specific area such as industrial area the PM$_{10}$ level was higher that the residential area. The PM10 levels in weekdays or during school hours were higher than in weekends due to the traffic activity. Some studies also showed that in several schools the PM$_{10}$ concentration has exceeded the maximum standard of PM$_{10}$ level. The higher PM$_{10}$ level has affected the children respiratory and has caused the children to be more prone to asthma.

Keywords: IAQ, PM$_{10}$, daycare centre, pollutants, Malaysia

1. Introduction

Indoor air quality in daycare centre can have a substantial impact on children's health, as an important environment where children may be exposed to pollutants and allergens. On average, children spend about 90 percent or more of their time indoors. That is one of the reasons that indoor air quality plays a major role in children's health (Zhang et al., 2006).

Air pollution is defined as the presence of undesirable levels of physical or chemical impurities. Air pollution, both indoor and outdoor, is often considered the major cause of environmental health problems for human. Even few years back, the problems associated with outdoor pollution have been well publicized due to the prominence of major pollutant sources (e.g., traffic, industrial, construction, combustion sources, etc.). However, in recent years, public concerns on indoor air quality (IAQ) have drawn a great deal attention, as the isolation of indoor from outdoor environment become phenomenal with the widespread supply of tight-sealed buildings and associated sick building syndrome (SBS) (Kabir et al., 2011). It is now known that indoor air pollution (IAP) likely has equal or even greater impact on human health when compares to that of outdoor air pollution. This happens because time spent indoor is usually higher that time spent outdoor, also the levels of some specific indoor pollutant, such as volatile organic compounds and particulate matter (PM), frequently exceed those found outdoor (Sousa et al., 2012).

Many organization such as the World Health Organization (WHO, 1999) recognized particulate matter (PM), carbon monoxide (CO), nitrogen dioxide (NO_2), ozone(O_3), lead (Pb) and sulfur dioxide (SO_2), as classical pollutants presenting a hazard to sensitive populations (Sherman, 2003).

2. Particulate Matter (PM_{10})

Particulate matter (PM_{10}) is used to describe particles with diameter of 10 micrometers or less for solid or liquid found suspended in the atmosphere. While individual particles cannot be seen with the naked eye, collectively they can appear as black soot, dust cloud or grey hazes. Particulate matter may be generated by natural process (e.g., pollen, bacteria, viruses, fungi, mold, yeast, salt spray, soil from erosion) or through human activities, including diesel trucks, power plants, wood stoves, and industrial processes. Particulate matter can be directly emitted or can be formed in the atmosphere when gaseous pollutants such as SO_2 and NO_X react to form fine particles (Aleesha et al., 2011).

Air particulate matter pollution is a major problem in most parts of Asia. The annual averaged values of total suspended particulate (TSP) in many cities frequently exceed 300 µg

m^{-3} (Baldasano et al., 2003). In Malaysia PM_{10} is one of the major air pollutants and is decisive in the computation of Malaysian Air Pollution Index (MAPI) (Afroz, 2003). PM_{10} concentrations in ambient air in Malaysia are monitored based on Recommended Malaysian Guidelines (RMG) at a threshold of 150 µg m^{-3} for 24 h average and an annual means of 50 µg m^{-3}. The air pollution in Malaysia is reported in the form of index, which allows the general public to better understand the meaning (Ramli, 2001). Furthermore, the concentration of air particulate matter is influenced by the southwest monsoon wind and the occurrence of biomass burning (Abas et al., 2004). During the normal period (without haze) the level of particulate matter is mostly influenced by motor vehicles and industry (Afroz et al., 2003). The level of PM_{10} tends to be lower during the rainy October until November period. This is partially due to particulate washout effects and the absence of strong ground-based inversions (Awang et al., 2000).

3. PM_{10} in Daycare Centres

The indoor air quality in the classroom of schools or daycare centres is primary concern as children spend a substantial amount of their time in such indoor spaces and they belong to more sensitive group of population. PM_{10} for indoor and outdoor air for a wind-induced natural ventilated airspace depended strongly on the ambient particles distribution and the design of the building openings (Liao et al., 2003). PM_{10} concentrations are related to their ambient outdoor concentrations, building characteristics, ventilation rate, air leakages, infiltration efficiency, and air mixing factor (Hinds, 1999).

The objective of this paper is to review the available studies that have been done by previous researchers concerning of indoor and outdoor PM_{10} concentrations in daycare centre and schools. This review is to provide a summary of the available scientific finding concerning the concentration, indoor/outdoor ratios and the sources.

4. Methodology

This review was to elaborated report of the original research studies performed on the measurement of PM_{10} in the children's environment. The literature research was conducted using the on-line database Science Direct, Scopus, and PubMed. The search performed, 25 articles containing data on PM_{10}, which are discussed below.

5. Results

The report studies of PM_{10} concentration in daycare centre and schools have been found from many countries such as Malaysia, Thailand, Hong Kong, India, Germany and Netherland from year 1997 to 2012. The review focused on the study in Asian Country due to the similarity of climate and weather. Table 1 shows many types of PM_{10} detectors that have been used from previous researcher. According to Ismail et al., 2010; Janssen et al., 2012; Lee and Chang, 2000 the equipment should be placed at 0.6 to 1.5 m above ground level at indoor and outdoor locations. Most PM_{10} concentration at daycare centre and schools has been taken in duration of 24 hours.

Table 1. Type of some of equipment was used from some previous researchers

No.	Researchers	Country	Equipment	Time Frame	Location of School
1.	Zailina et al. (1997)	Malaysia	The micro computer system of air monitoring unit 2 (MCSAM)	24 hours hourly in 4 months duration, data have been recorded every hourly	Sentul, Segambut, Titi Wangsa, Setapak, Kuala Lumpur
2.	Lee and Chang (2000)	Hong Kong	TSI DUSTTRAK	24 hours	Residential, industrial and rural area
3.	Abdul Mujid et al.(2003)	Malaysia	Air Sampling Pump	24 hours	Sungai Siput, Perak
4.	Rashida (2005)	Malaysia	Buck Ginie air sampling pump	24 hours	Kota Bahru Kelantan
5.	Fromme et al. (2007)	Germany	Optical laser aerosol spectrometer (LAS) (Dust Monitor 1.108, Grimm Technologies, Germany)	School hours (5hours), data have been recorded every minute.	City, rural and small town in Munich
6.	Marzuki et al. (2010)	Malaysia	Casella Microdust Pro	8:00 am to 1:00pm, every 5 minutes data have been recorded (1 month)	Vernacular school in Kuala Terengganu
7.	Pudpong et al. (2011)	Thailand	TSI DUSTTRAK Aerosol Monitor model-8520	24 hours (1.5 months)	High polluted and low polluted

8.	Latif et al. (2011)	Malaysia	Atomic absorption spectrophotometer with a graphite furnace	24 hours	Selangor
9.	Goyal and Khare (2011)	India	Environmental dust monitors model-107	School hours from 8:00am to 2:00pm, every 1 minute data have been recorded (7months)	Near a heavy-traffic road
10.	Janssen et al. (2012)	Netherland	Harvard Impactor	School hours (8hours), duration 1.5 months	Schools in inner city of Amsterdam

Zailina et al. (1997) has studied the relationship between air pollutants (PM_{10}) with asthmatic attacks in Kuala Lumpur areas while Mujid et al. (2003) and Rashida (2005) have done studies on the association between PM_{10} concentrations and the children respiratory system of school children in Perak and Kelantan, Malaysia. Marzuki et al. (2010) and Latif et al. (2011) compared the concentration of PM_{10} of schools and preschools in Kuala Terengganu and Selangor, Malaysia respectively.

Pudpong et al. (2011) selected 50 daycare centres from 15 districts in Bangkok, 25 centres from the high polluted area and 25 centres from the low polluted area. Lee and Chang (2000) also investigated PM_{10} concentration in five classrooms of residential, industrial and rural areas in Hong Kong. In India, Goyal and Khare (2011) assessed indoor air quality modelling for PM_{10} in classrooms of school building due to the concern on its potential effects on students health and performance.

In Europian studies, Fromme et al. (2007) and Janssen et al. (2012) have studied the sources of PM_{10} in the classrooms and its association with health risk and the correlation between classroom and outdoor concentrations of mass and elements of PM_{10} in Munich, Germany and Amsterdam, Netherland respectively.

Pudpong et al (2011) has reported that there was no significant difference of indoor and outdoor concentrations of PM_{10} in between the high and low polluted area. The concentration of PM_{10} in the high polluted area and the low polluted area are about 75.08 µg m^{-3} and 65.1 µg m^{-3} respectively. Both indoor and outdoor PM_{10} average levels measured at five classrooms in Hong Kong has exceeded the annual standard which ranged from 21 – 617 µg m[-3] with the maximum indoor PM_{10} level exceeded 1000 µg m[-3]. Marzuki et al. (2010) has also found that in one of the schools in Kuala Terengganu, Malaysia, the average PM_{10} concentrations was 194.3 µg m[-3] which has exceeded the recommended threshold level for PM_{10} in Malaysian Code of Practice (DOSH, 2005) of 150 µg m[-3]. The result obtained from Latif et al. (2011) found that the PM_{10} concentrations in the indoor environment was higher in schools (118.85 ± 86.08 ng.[m-3]) than in preschool (6 ± 0.30 ng.[m-3]).

Goyal and Khare (2011) reported that indoor average mass concentrations of PM_{10} are higher than their outdoor average mass concentration at every time interval during weekdays and weekends. In the classroom, PM_{10} during school hours has been shown to be 2-5 times higher than outdoor concentrations and twice as high as the average 24 h classroom (Jansen et al., 1999; Roorda-Knape, 1998), based on comparison of measurements from personal samplers and fixed-site monitors suggested that long-term childhood exposure to PM_{10} was on average three times higher than that indicated by outdoor measurements alone. Fromme et al. (2007) found that higher concentrations of PM_{10} were observed in winter than in summer. The PM10 fluctuated between 16.3 and 313 µg m[-3] while in summer the measurements ranged between 18.3 and 178 µg m[-3]

Diapouli et al. (2007) has investigated indoor/outdoor of PM_{10} and $PM_{2.5}$ concentration in seven primary schools in the Athens during the winter of 2003-2004 and 2004-2005. Mean of PM_{10} concentration was 236.13 µg m[-3] indoors and 162.89 µg m[-3] outdoors. The corresponding mean values for $PM_{2.5}$ were 82.65 µg m[-3] indoors and 56.25 µg m[-3] outdoors respectively. Indoor PM levels were greatly influenced by the presence of people and intensity of indoor activities. The variation in indoor PM concentration cannot be explained as functions of outdoor PM concentrations. Particle resuspension by normal activities of building occupants is an important factor in indoor particle concentrations. Thatcher and Layton (1995) found that light activities by four people, or continuous walking and sitting by a single occupant, caused a 2-4 times increase in the concentration of $PM_{>5<10}$.

Major roadways have been shown to be a significant source of PM_{10} air pollution. PM_{10} air pollution has been demonstrated to be an important factor associated with negative respiratory health effects, especially in minority children. The result was found by Korenstein & Piazza (2002), students in close proximity to major roadways receive a dose of PM_{10} at levels approaching 10-15 micrograms per cubic meter, an exposure predicted to cause negative health effects. Similar study has been done by Ekmekciouglu and Keskin (2007), the PM_{10} and $PM_{2.5}$ maximum concentration limits determined by World Health Organization were exceeded considerably in four schools located nearby a major road with high traffic density. Zailina et al. (1997) has monitored that during haze period the daily mean concentrations of PM10 was 83.87 $\mu g\ m^{-3}$, while the maximum value of 285 $\mu g\ m^{-3}$ was nearly twice the Malaysian guidelines of 150 $\mu g\ m^{-3}$ which can cause reduced visibility and influenced the respiratory functions of children as they are very vulnerable to these pollutions which may trigger asthmatic attacks. Abdul Mujid et al. (2003) has carried out a cross sectional study and found that PM_{10} concentrations showed a significantly higher mean of 76.66 $\mu g\ m^{-3}$ in exposed to quarry dust areas than the comparative areas.

The PM_{10} concentration level in different areas has been done by Kim et al (2005). The metropolitan city is in the central of city and suburb area in the Korea has show there are no significant difference of PM_{10} level after recorded one month (55.3 vs.52.3$\mu g\ m^{-3}$). In both regions, the difference of PM_{10} level between March and December was statically significant (64 vs.56$\mu g\ m^{-3}$) in the city and 64 and 54 $\mu g\ m^{-3}$ in the surburb. In the contrary result that had been found by Simons et al. (2007), they compared the indoor concentration of PM_{10} level in the inner city and suburban homes. The indoor of PM_{10} in inner city was 47 $\mu g\ m^{-3}$ and for the PM_{10} level in suburban is 18 $\mu g\ m^{-3}$.

Another study, Sulaiman et al. (2005) investigate concentration and composition of PM_{10} in outdoor and indoor air in industrial area of Balakong Selangor, Malaysia. The concentrations of indoor air was found to be higher than those of the outdoors in the study area ($p<0.05$). The highest concentration of PM_{10} in the indoor air is 149 $\pm$ 14.73 $\mu g\ m^{-3}$and the highest concentration of PM_{10} of 128 $\pm$ 34.37 $\mu g\ m^{-3}$in the outdoor respectively. Differences in the PM_{10} concentrations for both indoor and outdoor were likely caused by different activities occurring at the two stations. Also Li (1994) investigated PM_{10} level of three home in different area. The indoor of TSP levels ranged 20 to 70 $\mu g\ m^{-3}$ for home (1st floor) in the residential area, 60 to 350 $\mu g\ m^{-3}$ for home (1st floor) in the industrial area and 30 to 200 μg m^{-3} for home (5th floor) in industrial area. The higher particle mass concentration of homes 2 and 3 are probably related to their surrounding environments. eg. some manufacturing foundries in the industrial area. Moreover, result obtained by Diapouli et al. (2008) showed that the concentration of PM_{10} level for homes at first floor of an apartment building in residential area in weekend is 68.2 μg m^{-3}and 96.3 μg m^{-3} in weekdays. The lower concentration PM_{10} in weekends compare to weekdays possibly due to the increased traffic density during working days.

6. Conclusions

The reviews have concluded that the indoor PM_{10} concentrations are higher compare to outdoor PM_{10} concentrations concentration in schools. Higher number of children in the rooms may contribute to higher indoor concentrations. Age of building, types of flooring, curtains, dust from blackboard and shelf area and fans were found to be determinant of PM_{10} in classroom (Marzuki et.al. 2010). Schools with locations near industrial areas and main roads were proofed to have higher concentration of PM_{10}. High in PM_{10} concentration level was proven to have caused respiratory problems on children.

Reference

1. Abas, M.R.B., Oros, D.R., and Simoneit, B.R.T. 2004. Biomass burning as the main source of organic aerosol particulate matter in Malaysia during haze episodes. Chemosphere 55, 1089-1095.

2. Abdul Mujid, A. Zailina, H., Juliana, J., Shamsul Bahri, M. T. 2003. Malaysian Journal of Public Health Medicine 3(2):23-32.

3. Afroz, R., Hassan, M.N., and Ibrahim, N.A. 2003. Review of air pollution and health impact in Malaysia. Environmental Research 92, 71-77.

4. Aleesha, N.A., Shuhaimi, S.H., Ying, T,Y. Shapee. A.H., and Maznorizan, M. 2011. The study of seasonal of PM_{10} concentration in Peninsular, Sabah and Sarawak. Malaysian Meteorological Department, Ministry of Science, Technology and Innovation, Vol.9, 1-27.

5. Baldasano, J.M., Valera, E., and Jimenez, P. 2003. Air quality data from large cities. The Science of Total Environment, 307, 141-165.

6. Diapouli, E., Chaloulakou, A., and Spyrellis, N. 2007. Indoor and Outdoor particulate matter concentrations at schools in the Athens area. Indoor and Built Environment 16, 155-161.

7. Diapouli, E., Chaloulakou, A., and Spyrellis, N. 2008. Indoor and outdoor PM concentrations at residential environment, in the Athens area. Global NEST Journal 10, 201-208.

8. Ekmekciouglu, D and Keskin, S.S. 2007. Characterization of indoor air particulate matter in selected elementary schools in Istanbul, Turkey. Indoor and Built Environment 16, 169-176.

9. Engku Aminatul Rashida Binti Engku Ariff, 2005. Pendedahan Kepada Pencemaran Udara Dalaman (PM_{10}) Dan Hubungannya Dengan Kesihatan Respiratori Di Kalangan Kota Bahru, Kelantan. Thesis. Fakulti Perubatan dan Science Kesihatan, Universiti Putra Malaysia.

10. Fromme, H., Twardelle, D., Dietrich, S., Heitmann, D., Schierl, R., Liebl, B., Ruden, H. 2007.Particulate Matter in The Indoor air of Classroom- Exploratory results from Munich and Surrounding Area. Atmospheric Environment 41, 854-866.

11. Hinds, W.C. 1999. Aerosol technology: properties, behaviour, and measurement of airborne particle. New York: Willey.

12. Janssen, N., Hoek, G., Harssema, H., and Brunekreef, B. 1999. Personal exposure to fine particles in children correlates closely with ambient fine particles. Archives of Environmental Health 54, 95-101.

13. Juliana, J., Dayang, A.A., Jamal, H.H., Zailina, H., Bilkis, A. Z. 2001. Influence of Indoor Air Pollution on Respiratory Illness and Lung Function Among Children in Hulu Langat. Community Health Journal 7, 51-56.

14. Kabir, E., Kim, K.H., Sohn, J.Y., Kwoon, B.Y., and Shin, J.H. 2011. Indoor air quality assessment in child care and medical facilities in Korea. Environment Monitor Assessment, DOI 10.1007/s10661-011-2428-5. Springer.

15. Kim, J.H., Lim, D.H., Jeong, S.J., and Son, B.K. 2005. Effects of particulate matter (PM10) on the pulmonary function of middle-school children. Journal Korean Medical Science 20, 42-45.

16. Korenstein, S and Piazza, B. 2002. An exposure assessment of PM_{10} from a major highway interchange: are children in nearby schools at risk?. Journal Environment 65, 17-37.

17. Latif, M.T., Baharudin, N.H., Nor, Z.M., and Mokhtar, M. 2011. Lead in PM_{10} and Indoor dust around schools and preschools in Selangor, Malaysia. Indoor and Built Environment 20, 346-353.

18. Lee, S.C. and Chang, M. 2000. Indoor and Outdoor Air Quality Investigation at Schools in Hong Kong. Chemosphere 41, 109-113.

19. Li, C.S. 1994. Relationship of indoor/outdoor inhalable and respirable particles in domestic environments. Science Total Environment 151, 205-211.

20. Liao, C.M., Huang, S.J., and Yu, H. 2003. Size-dependent particulate matter indoor/outdoor relationship for a wind-induced naturaly ventilated airspace. Building and Environment 39, 411-420.

21. Marzuki, I, Nur Zafirah, M. S., Ahmad Makmon, A. 2010. Indoor Air Quality in Selected Samples of Primary Schools in Kuala Terengganu, Malaysia. Environmental Asia 3, 103-108.

22. Pudpong, N., Rumchev, K., and Kungkuliti, N. 2011. Indoor concentration of PM_{10} and factora influencing its concentration in daycare centres in Bangkok, Thailand. Asia Journal of Public Health 2, 3-12.

23. Ramli. 2001. Seasonal locational and diurnal variations of particulate matter in small towns in Wales and Malaysia. University of Wales Aberystwyth. United Kingdom.

24. Roorda-Knape. 1998. Air pollution from traffic in city districts near major motorways. Atmospheric Environment 32, 1921-1930.

25. Sherman, M.H. and Matson, N.E. 2003 Reducing indoor residential exposure to outdoor pollutans. AIVC Technical Note 58, 1-32.

26. Sions, E., Brosnan, J.C., Buckley, T., Breysse, P., and Eggleston, P.A. Indoor environmental differences between inner city and suburban home of children with asthma.

27. Sousa S.I.V., Alvim-Ferraz, M.CM. and Martins, F.G. 2012. Indoor air PM_{10} and $PM_{2.5}$ at nurseries and primary schools. Advanced Materials Research 433-40, 385-390.

28. Sulaiman, N, Abdullah, M., and Chieu, P.L.P. 2005. Concentration and composition of PM_{10} in outdoor and indoor air in industrial area of Balakong Selangor, Malaysia. Sains Malaysiana 34, 43-47.

29. Zailina, H., Juliana, J., Norzila, M.Z., Azizi, H.O., Jamal, H. H. 1997. The Relationship Between Kuala Lumpur Haze and Asthmatic Attacks in Children. Malaysian Journal of Child Health 9, 151-159.

30. Zhang, G., Spickett, J., Rumchev, K., Lee, A.H., and Stick, S. 2006. Indoor "environmental quality in a low allergen" school and three standard primary schools in Western Australia". Indoor Air 16, 74-80.